Illustrated Genera of
Rust Fungi

Revised Edition

George B. Cummins
   Department of Plant Pathology
   The University of Arizona, Tucson

Yasuyuki Hiratsuka
   Northern Forest Research Centre
   Canadian Forest Service, Edmonton

Published by
The American Phytopathological Society
St. Paul, Minnesota

This book has been reproduced directly from typewritten copy submitted in final form to The American Phytopathological Society by the authors of the volume. No editing or proofreading has been done by the Society.

To facilitate a reduced purchase price, the authors have waived royalties.

Library of Congress Catalog Card Number: 83-72397
International Standard Book Number: 0-89054-058-6

Printed in the United States of America

The American Phytopathological Society
3340 Pilot Knob Road
St. Paul, Minnesota 55121, USA

PREFACE

This manual is designed primarily to be useful in the identification of rust fungi and to introduce students to the spore states, life cycles, and morphological diversity of the Uredinales. If the sequence of the 105 genera stimulates thought and produces ideas concerning possible relationships that will be good. The sequence suggests our concept of relationships, as does the assignment of the genera to 14 families. An alternative arrangement of the genera could have been alphabetical and this doubtless would have pleased some persons, especially those concerned with specific genera. But, as an aid to identification, grouping by morphological similarity is more useful. Possibly, one will need only to look at the illustrations to reach at least a tentative identification. In fact, much can be learned about the Uredinales just by looking at the pictures!

There is not now, and probably never will be, unanimity of opinion concerning the numbers of genera and the characters that delimit them. This is because different emphasis is accorded to the various structures and to the importance of the relationship of the host plants parasitized. Although this situation may be confusing, it is not undesirable because it stimulates discussion and additional study. There have been four previous descriptive treatments of the genera of the Uredinales. In 1900, Dietel (Natürliche Pflanzenfamilien, Teil I), recognized 33 genera, but in 1928 (Natür. Pflanzenfam., Band 6) he included 101 genera. The increased numbers resulted largely from new studies in various regions of the world. Thirumalachar and Mundkur (Indian Phytopathology 1949-1951) described and illustrated 122 genera and, in 1959 (Illustrated Genera of Rust Fungi), Cummins recognised 95 genera. The differing numbers reflect different opinions as much or more than changes in numbers of genera.

Because generic identity usually is only a first goal, we have listed pertinent literature following the generic description. This literature may provide means to identify species, or may refer to macroscopic illustrations or to electron micrographs of structures (TEM) or surface ornamentation (SEM). But there are genera which have so few species (sometimes only one) or that are so little known that there is no special literature. Also, following the section devoted to genera, there are listed descriptive manuals that treat the Uredinales of various countries or of certain groups of host plants.

CREDITS AND ACKNOWLEDGMENTS

Figure 1 is a rearrangement of figures 1-11, in part, of Hiratsuka and Sato in K.J. Scott and Chakravorty, eds., The Rust Fungi, 1982, copyright Academic Press, used by permission; the drawings originally from Hiratsuka and Hiratsuka, 1980, and most of the photomicrographs of spermogonia from Hiratsuka and Cummins, 1963. We acknowledge the helpful suggestions of Roger S. Peterson, St. John's College, Santa Fe, NM, relative to the genera of rust fungi of importance in forest pathology. Joe F. Hennen, Purdue University, West Lafayette, IN and Pablo Buriticá, Colombian Agricultural Institute, Bogota, discussed, in person with the senior author, the entire section devoted to generic descriptions. This contribution of ideas and time is very gratefully acknowledged.

TABLE OF CONTENTS

Illustrated Genera of
Rust Fungi
Revised Edition

INTRODUCTION

WHAT ARE THE RUST FUNGI

The rust fungi are a unique and interesting group of fungi belonging in the order Uredinales of the class Basidiomycetes, which implies that in their reproductive cycle they produce basidia and basidiospores. The name "rust fungi" comes from the fact that often pustules and spores have a rusty yellow color. Although rust fungi are not as conspicuous as mushrooms, they are not difficult to recognize once a person has been introduced to a few of them. In their natural habitat rust fungi are obligate parasites of living plants, although a few species are now cultured on artificial media. Rust fungi parasitize a wide range of host plants, including ferns, conifers, and angiosperms (both mono- and dicotyledonous). About 5000 species have been recognized and some 300 generic names have been proposed (Laundon 1965) of which between 100 and 125 are recognized as "good" genera (Cummins 1959; Thirumalachar and Mundkur 1949, 1950; Hiratsuka 1955). In this manual 105 genera are described and illustrated, plus four endocyclic "genera."

Rust fungi of Europe, North America, Japan, New Zealand, and Australia have been reasonably well catalogued, but significant numbers of new species and perhaps genera are still to be expected in tropical and subtropical regions of South America, Africa, and southeast Asia.

The rust fungi have three unique features: 1) up to five morphologically different spore states exist in a life cycle of a single species; 2) many species need two unrelated groups of host plants to complete their life cycle (heteroecious life cycle), although others can complete the life cycle on a single kind of host plant (autoecious life cycle); and 3) usually species have narrow and specific host ranges.

ECONOMIC IMPORTANCE OF THE RUST FUNGI

Rust fungi as a group are among the most economically important pathogens of many important native and cultivated plants. Unlike other kinds of plant pathogens, which tend to attack weakend or poorly growing plants, rust fungi parasitize fresh tissues of vigorously growing plants. For this reason, problems with rust fungi tend to increase with intensive and extensive cultivation of many economically important agricultural, horticultural, and forest crops. These fungi are the major concerns and limiting factors for successful cultivation of such internationally important crops as wheat, corn, coffee, and pine.

Most textbooks of plant pathology describe and illustrate such rust diseases as wheat stem rust (Puccinia graminis Pers.) and white pine blister rust (Cronartium ribicola J.C. Fisch. ex Raben.) as examples of economically important rust diseases. Examples of other diseases caused by rust fungi are:

Wheat leaf rust (Puccinia recondita Rob. ex Desm.)
Wheat stripe rust (Puccinia striiformis West.)
Corn leaf rusts (Puccinia sorghi Schw., P. polysora Underw.)
Flax rust (Melampsora lini (Pers.) Lév.)
Coffee leaf rust (Hemileia vastatrix Berk. & Br.)
Mint rust (Puccinia menthae Pers.)
Sugarcane rusts (Puccinia kuehnii Butl., P. melanocephala Syd.)
Hard pine stem rusts (Cronartium flaccidum (Alb. & Schw.) Wint., C. comandrae Pk.,
   C. comptoniae Arth., C. coleosporioides Arth., C. quercuum Shirae ex Miyabe
   (incl. C fusiforme Hedgc. & Hunt), Endocronartium harknessii (J.P. Moore) Y.
   Hirat., E. pini (Lév. emend Kleb.) Y. Hirat.)
Chickpea rust (Uromyces ciceris-arietinus Jacz.)
Cotton rust (Puccinia cacabata Arth. & Holw.)
Bean rust (Uromyces appendiculatus (Pers.) Unger)

Peanut rust (<u>Puccinia</u> <u>arachidis</u> Speg.)
Sunflower rust (<u>Puccinia</u> <u>helianthi</u> Schw.)
Mulberry rust (<u>Aecidium</u> <u>mori</u> Barcl.)
Poplar and willow leaf rusts (<u>Melampsora</u> spp.)
Pine needle rusts (<u>Coleosporium</u> spp.)
Spruce needle rusts (<u>Chrysomyxa</u> spp.)
Cedar-apple rust (<u>Gymnosporangium</u> <u>juniperi-virginianae</u> Schw.)
Eucalyptus rust (<u>Puccinia</u> <u>psidii</u> Wint.)
Onion rust (<u>Puccinia</u> <u>allii</u> Rud.)
Peach and plum rust (<u>Tranzschelia</u> <u>discolor</u> (Fckl.) Tranz. & Litv.

By using the narrow and specific host ranges of rust fungi, attempts have been made to control certain weeds. Examples are 1) rush skeletonweed (<u>Chondrilla</u> <u>juncea</u> L.) by <u>Puccinia</u> <u>chondrillae</u> Corda in Australia (Hasan and Walshore 1973), 2) blackberry (<u>Rubus</u> spp. by <u>Phragmidium</u> <u>violaceum</u> (Schul.) Wint. in Chile (Oehrens 1977), and 3) water hyacinth (<u>Eichhornia</u> sp.) by <u>Uredo</u> <u>eichhorniae</u> Gonz-Frag. & Cif. in the southern United States (Charudattan et al. 1976).

THE SPORE STATES OF THE RUST FUNGI

Five kinds of spore producing structures are generally recognized as basic spore states of the rust fungi, as follows: spermogonium (pl. spermogonia); aecium (pl. aecia); uredinium (pl. uredinia) telium (pl. telia); and basidium (pl. basidia). They are designated often by Roman numerals as 0, I, II, III, and IV, respectively.

Two different bases have been applied in the definitions and terminology of spore states. They are the "morphologic" system and the "ontogenic" system. The morphologic system, as defined by Laundon (1967) and Holm (1973), emphasizes the morphology of spores as the basis for defining spore states. In this system, aeciospores are defined as produced in chains and usually having ornamentation traditionally known as verrucose and urediniospores are defined as being always unicellular and borne singly on pedicels, and usually having ornamentation traditionally known as echinulate (Laundon 1967).

On the other hand the ontogenic system, used by Cummins (1959), following Arthur (1905, 1925, 1929) and expanded by Hiratsuka (1973b, 1975) emphasizes positions of the spore states in the life cycle rather than as clearly recognizable morphological entities as the basis for spore terminology. Definitions of spore states according to the ontogenic system (Hiratsuka 1973b) are as follows:

Teliospores: basidia-producing spores (probasidia; hypobasidia).
Basidiospores: monokaryotic spores produced on a basidium.
Spermatia: male gametes.
Aeciospores: non-repeating spores produced as a result of dikaryotization, thus in sori normally associated with spermogonia, and which give rise to dikaryotic vegtative mycelium.
Urediniospores: repeating vegetative spores produced on a dikaryotic mycelium.

Terminologies based on either system are the same for many well known species such as <u>Puccinia</u> <u>graminis</u>, <u>P</u>. <u>helianthi</u>, <u>Cronartium</u> <u>ribicola</u>, and <u>Melampsora</u> <u>lini</u> but differ in other situations. For example, urediniospores (by the ontogenic system) of <u>Chrysomyxa</u> and <u>Coleosporium</u>, which are produced on non-coniferous alternate hosts, are produced in chains and have verrucose ornamentation and should be called aeciospores by the morphologic system. These genera would, then , have two aeciospore states, one on the aecial host and one on the telial host. Also, it is impossible to define teliospores by their morphology because so much variation exists among the various genera. Thus, the only possible definition is 'the spore state that gives rise to basidia', which is the ontogenic definition. In this manual we use terminology based on the ontogenic system. For further discussion regarding these systems of terminology refer to papers by Laundon (1967, 1972), Hiratsuka (1973b, 1975), and Durrieu (1979).

SPERMOGONIUM

Spermogonia are produced on a haploid thallus which results from infection by a basidiospore. They are minute but because they occur in groups, often on discolored spots or hypertrophied tissues, they may be obvious macroscopically. Besides, the spermogonia typically are associated with visible aecia or telia. The spermogonia produce spermatia in a sweetish exudate and the spermatia function as sperms. Sperms are small, one-celled, hyaline spores that have little or no diagnostic value, but morphological types of spermogonia have been considered to

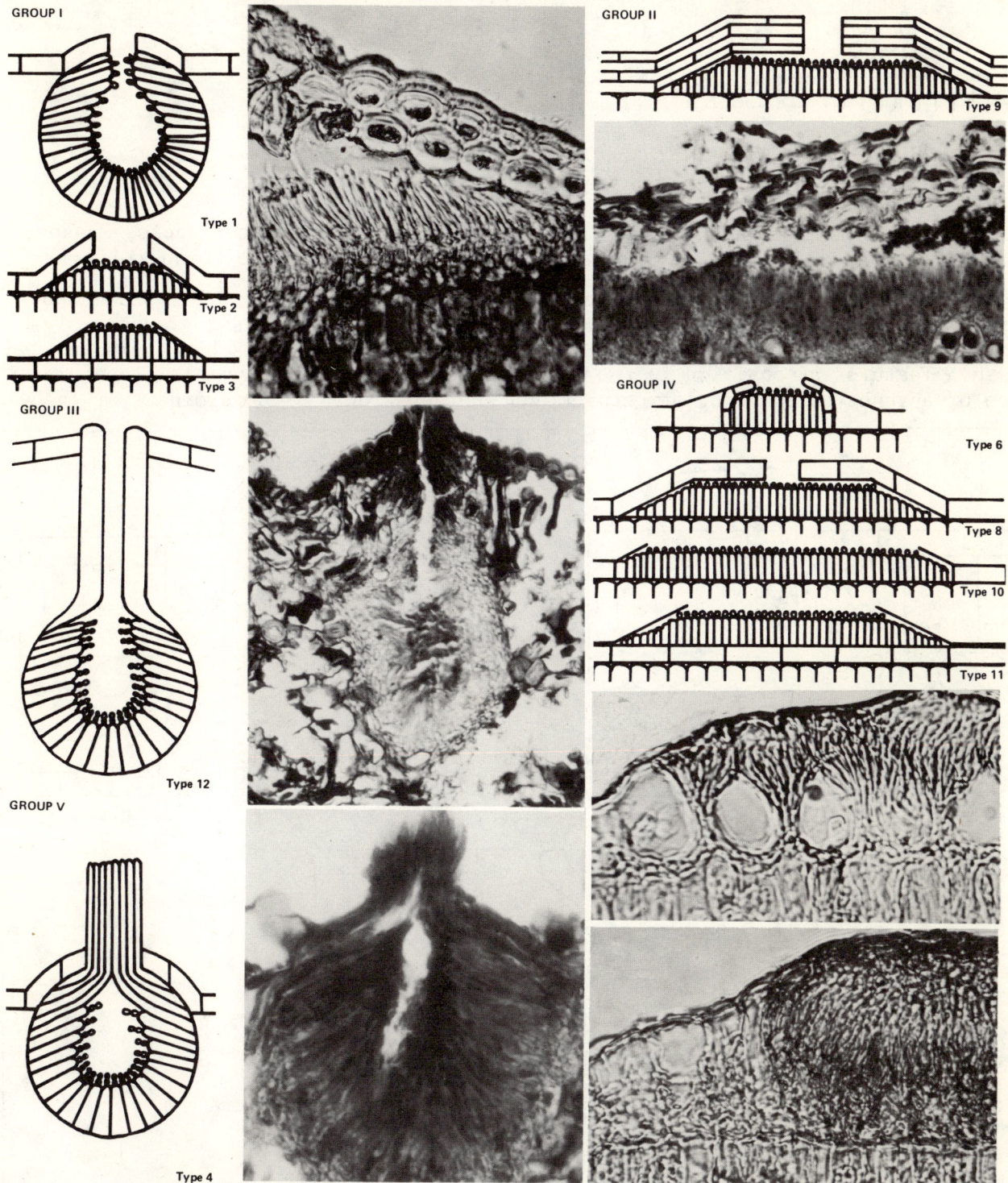

Fig. 1. Spermogonia. Group I, common in Pucciniastraceae. Group II, intracortical position of Cronartium. Group III, occurs in Mikronegeria. Group IV, characteristic of Phragmidiaceae. Group V, typical of Pucciniaceae and Pucciniosiraceae. (Used, by permission, from The Rust Fungi, ᶜAcademic Press). Continued next page.

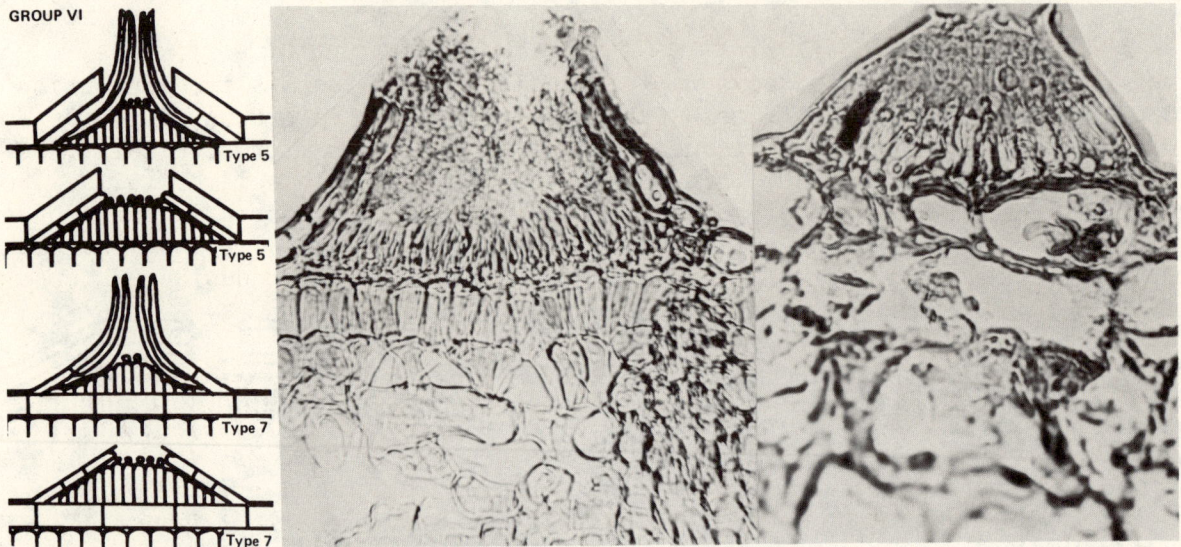

Fig. 1B. Group VI spermogonia. These types occur in the Chaconiaceae, Phakopsoraceae, Pileolariaceae, Raveneliaceae, Sphaerophragmiaceae, Uropyxidaceae.

be dependable as characteristics useful in suprageneric taxonomy (Hiratsuka and Cummins, 1963). Twelve morphological types were recognized and six groups designated (Fig. 1A, 1B) by Hiratsuka and Hiratsuka (1970). The types are characterized by such features as position in the host tissue, shape of the hymenial layer, presence or absence of bounding structures and whether growth is determinate or inde-

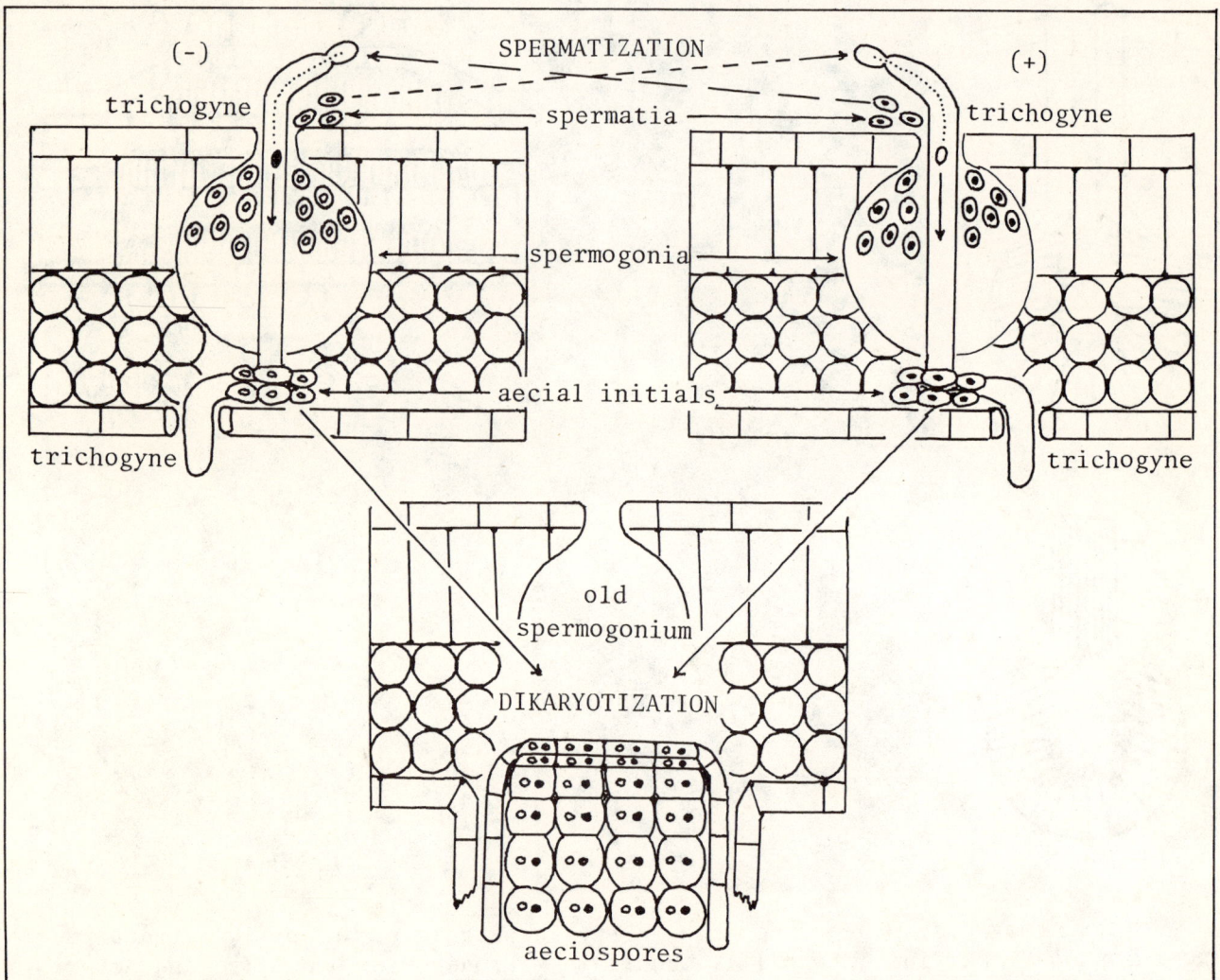

Fig. 2. Schematic representation of the sexual apparatus of the rust fungi.

terminate. Spermogonia may not develop and rarely are omitted from life cycles.

AECIUM.

Aecial initials (egg cells; female gametes) also are produced on a haploid thallus which results from the infection by a basidiospore. Aecia are preceded or accompanied by spermogonia, both structures arising on the same thallus. The aecial initials receive nuclei of opposite compatibility (= or -) from spermatia through trichogynes (flexuous or receptive hyphae) that extend through spermogonia or stomata from the egg cells and thus become dikaryotic (Fig. 2). After dikaryotization, aeciospores develop. Aeciospores are unicellular, and upon germination, produce dikaryotic mycelium which, in turn, produces either uredinia or telia but not aecia.

Aeciospore surface markings have been recognized most typically as "verrucose" but with close examination with SEM, at least eight distinct types of markings are recognized. They are; verrucose, aciculate, nailhead, echinulate, coronate, tubulate, annulate, and reticulate (Sato and Sato, 1982). In general, spore surface types are the same among species of the same genus. Besides regular fine warts on the spores, many species of Puccinia and Uromyces have structures called pore plugs or refractive granules. Savile (1973) recognized five types of aeciospores among Puccinia and Uromyces species attacking Cyperaceae, Juncaceae, and Poaceae based on the presence or absence of pore plugs and their distribution patterns.

1. Aecidium. Cup-shape structures with well-developed peridium (Fig. 3A) and with peridial cells more or less rhomboidal. Aeciospores are catenulate, mostly with intercallary cells, and usually with verrucose surface. The Aecidium is especially common in Puccinia and Uromyces but occurs in some other genera.

2. Peridermium. This kind of aecium has a well-developed peridium (Fig. 3B) with peridial cells that are relatively long and narrow. The aeciospores are catenulate and the surface sculpture often is more complex than simple warts (Fig. 5G). The peridermioid aecium is characteristic of genera such as Milesina, Cronartium, Coleosporium, and Pucciniastrum of the Melampsoraceae. The peridermioid type is actually neither morphologically homogeneous nor always distinctive because it intergrades with Aecidium. Peridermium traditionally is restricted to peridiate aecial states on gymnosperms.

3. Roestelia. The aecia of this type have well-developed peridia (Fig. 3C) and

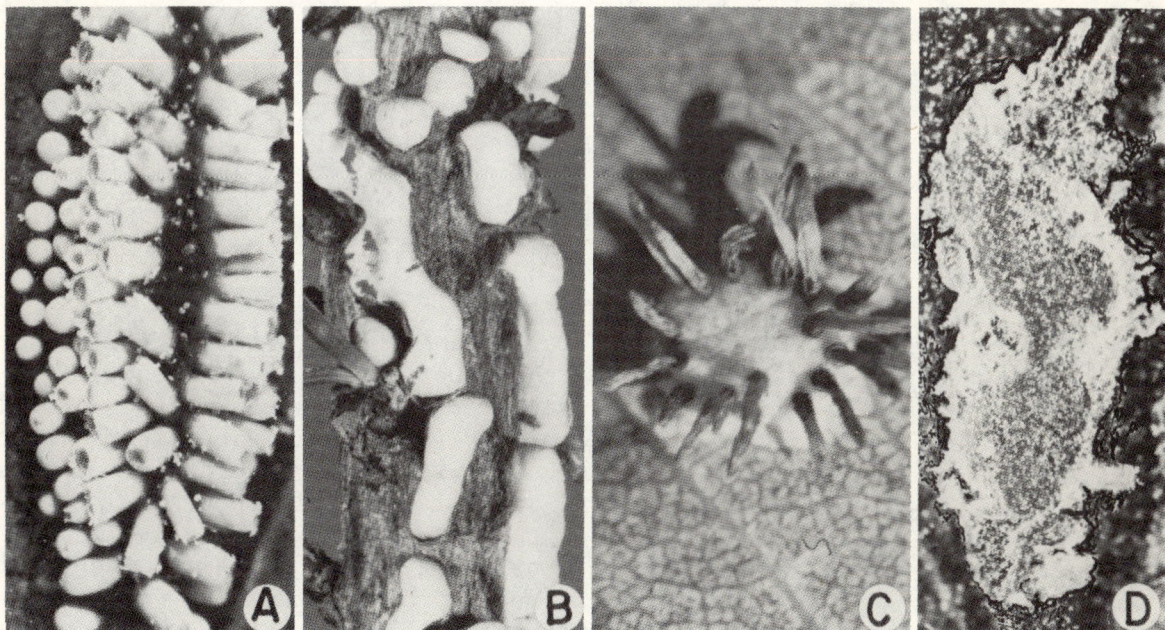

Fig. 3. Types of aecia. A. Aecidium, of Puccinia crandallii Pam. & Hume. B. Peridermium, of Cronartium coleosporioides Arth. C. Roestelia, of Gymnosporangium nelsonii Arth. D. Caeoma, of Phragmidium montivagum Arth.

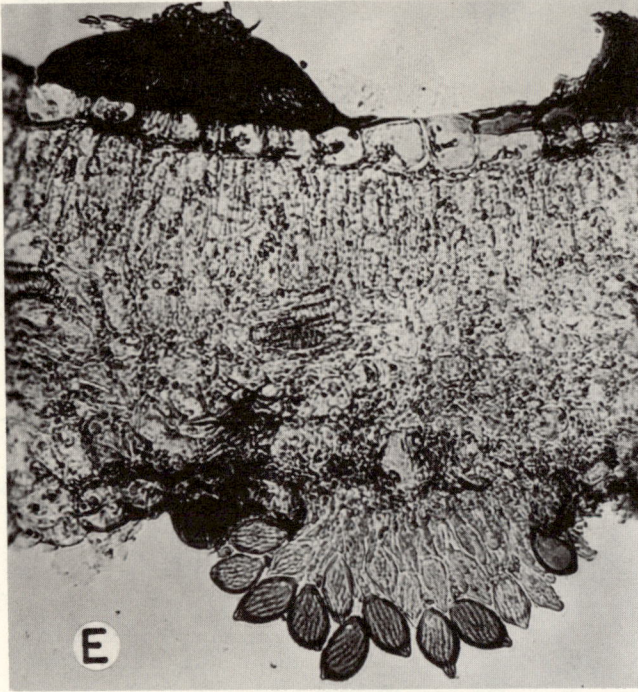

Fig. 3E. <u>Uraecium</u> of <u>Pileolaria</u> <u>brevipes</u>.

generally tend to shred longitudinally at maturity. The peridial cells are long and narrow. Aeciospores are catenulate and verrucose but not as deepsculptured as spores of <u>Peridermium</u>. The roestelioid aecium is relatively distinctive but it also intergrades with <u>Aecidium</u>. By general agreement, <u>Roestelia</u> applies only to the aecial states of <u>Gymnosporangium</u>. However, some species of <u>Puccinia</u> have strongly roestelioid aecia.

4. <u>Caeoma</u>. Aecia of this type have a rudimentary or no peridium (Fig. 3D) but some of them have paraphyses. Aeciospores are catenulate and mostly verrucose. Caeomatoid aecia are charateristic of such genera as <u>Melampsora</u> and <u>Phragmidium</u>. The distinction between <u>Aecidium</u> and <u>Caeoma</u> is not as clear as it sounds because some caeomatoid aecia of <u>Melampsora</u> are known to have rudimentary by distinct peridia.

5. <u>Uraecium</u>. This is the uredinoid aecium and probably the least understood or accepted. The pustules appear as uredinia but are accompanied by spermogonia and occupy the position of aecia in the life cycle. They produce spores only after spermatization. Uraecia have pedicellate , mostly echinulate aeciospores which are morphologically like the urediniospores of the species (Fig. 3E). In addition to being accompanied by spermogonia, uraecia are closely grouped, as is true of other aecial types, and not randomly scattered as are uredinia. Uraecia are characteristic of such genera as <u>Ravenelia</u>, <u>Prospodium</u>, <u>Uropyxis</u>, and <u>Pileolaria</u> but are common also in <u>Puccinia</u> and <u>Uromyces</u> among others. Aecia are useful in generic distinction but the aeciospores have been little used. Their use in delimiting species is greater but limited largely to size and surface sculpture. Germ pores are present but masked by surface sculpture. Scanning electron micrographs have clarified the nature of the sculpture and emphasize a need for more precise terms for surfaces usually blanketed by "rugose" and "verrucose". The verrucosity of aeciospores of <u>Puccinia</u> graminis has little in common with the verrucosity of aeciospores of <u>Cronartium</u> spp. and <u>Chrysomyxa</u> spp.

6. <u>Elateraecium</u>. These unique aecia are known to occur only in the genus <u>Hiratsukamyces</u>. The sorus is basically caeomatoid, having catenulate spores and no peridium. But among the spores are elater-like hyphae that expand out of the sorus carrying the spores with them

UREDINIUM.

Uredinia are produced by a dikaryotic mycelium and the urediniospores produce dikaryotic mycelium when they germinate. Thus this spore state is the "repeating state" of the rust fungi. The first crop of uredinia to be produced in the life cycle develops on mycelium which resulted from the germination of aeciospores, but the mycelium resulting from the germination of urediniospores may produce more uredinia but usually does ultimately give rise to telia and teliospores.

The uredinial sorus may vary greatly in morphology. It may have a well-developed peridium (e.g., <u>Milesina</u>, <u>Pucciniastrum</u>), or it may have paraphyses and no or obscure peridium (e.g., <u>Melampsora</u>), or it may be devoid of both (e.g., most species of <u>Puccinia</u>). There is another type of uredinium which deserves special attention. The uredinia look exactly as aecia but they occur scattered singly without accompanying spermogonia, thus meeting the requirements set forth for the uredinial state as to ontogeny and fate. They are, in fact, preceded in the life

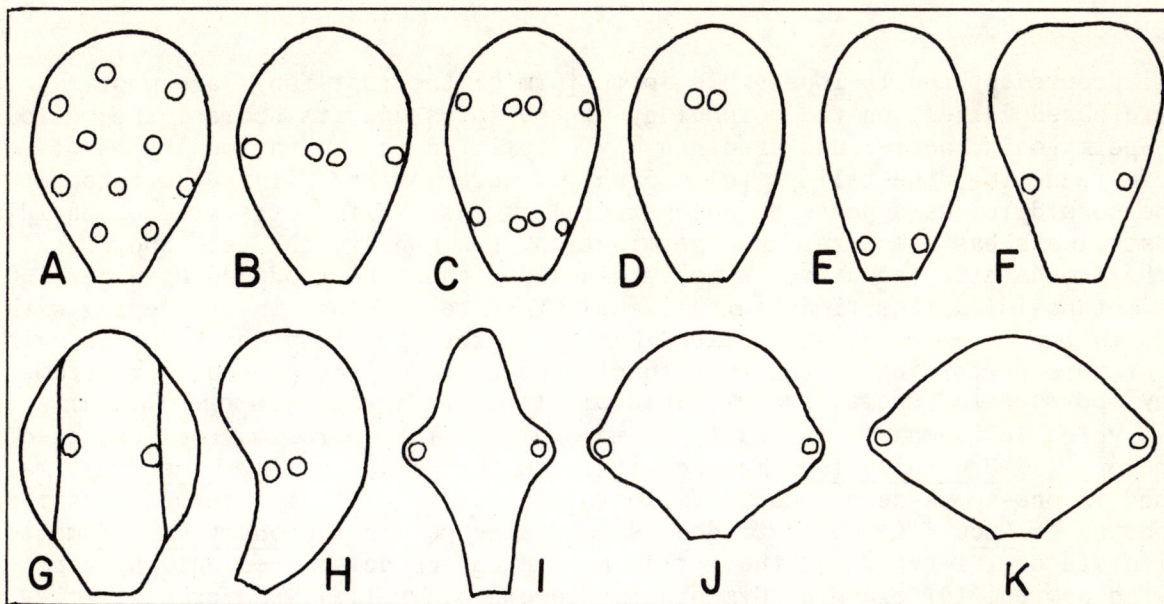

Fig. 4. Arrangements of germ pores and shapes of spores. A. Pores scattered; spores usually broadly ellipsoid or globoid. B. Pores equatorial, unizonate; spores usually obovoid or broadly ellispoid. C. Bizonate; spores usually tend to be ellipsoid. D. Superequatorial, unizonate. E. Basal, unizonate. F. super-equatorial, unizonate. G. Equatorial, pore caps exaggerated, uncommon. H. Equatorial, unizonate; spore reniform; common in certain legume rusts. I. Equatorial in lateral angles; spore obtrullate in outline; uncommon. J. Approximately equatorial; spore depressed globoid or helmet-shape. K. Equatorial, unizonate; spore transversely ellipsoid.

cycle by true aecia. This type often is called repeating aecium. Examining 111 species in 65 genera, Kenney (1970) recognized 14 morphological types of uredinia based on such characteristics as bounding structure, position of sori, and spore types. Sathe (1977) defined and illustrated 11 types. Urediniospores most common-ly are echinulate but may have other surface sculpture (Fig. 5). Almost all are one-celled but one exception has been reported in Gymnosporangium gaeumannii Zogg ssp. albertensis (Hiratsuka 1973), where occasional two-celled spores occur.

When only uredinial states are known, they are placed in the genus Uredo and a few other genera established as uredinial anamorphs. For the most part, uredin-iospores are capable of germination immediately upon formation and usually are short lived under natural conditions. But some species of rust fungi produce mod-ified urediniospores known as amphispores. Amphispores have thicker and usually more deeply pigmented walls and are capable of enduring extended periods of un-favorable conditions. Puccinia vexans Farl., Puccinia atrofusca (Dudl. & Thomp.) Holw. and Hyalopsora polypodii (Diet.) Magn. are species that produce amphispores as well as ordinary urediniospores.

Urediniospores, except as to mode of origin, i.e. catenulate or pedicellate, rarely figure in generic distinctions but are widely used to delimit species. shapes of spores and especially arrangements and numbers of germ pores are use-ful and constant characteristics (Cummins 1936). Most, and possibly all, uredin-iospores have germ pores but they may be indistinct or indiscernible, especially in spores with colorless or lightly pigmented walls. Pores may appear as clear spots, or may be detected because of small lens-like caps, or the wall may bulge inward slightly where pores occur. Numbers of pores range from one (uncommon) to at least 15. The principal arrangements are shown in Fig. 4, along with shapes of spores but this coincidence does not necessarily imply a direct relationship. Scattered and equatorial arrangements are the commonest. When pores are scatter-ed, a range of six to ten is common; when they are zonate, two to five are common. It is speculated that scattered and numerous germ pores represent primitive forms and equatorial and few germ pores represent advanced forms (Cummins 1936).

TELIUM.

Telia represent the indispensible spore form of the rust fungi and most genera are based largely on the morphology of the sorus and its spores, the teliospores. Spermogonia, aecia, and uredinia may be omitted in the various types of life cycle reduction; the telium (teleomorph) is never omitted if the rust concerned can be considered as a perfect (holomorphic) fungus. Teliospores produce basidia and almost always basidiospores upon germination. Frequently, the teliospore is called the probasidium (sometimes hypobasidium) and the tube produced upon germination the metabasidium (sometimes epibasidium). This terminology is in keeping with that used in the "higher basidiomycetes" but we prefer to retain the terms used in most literature pertaining to the rust fungi. The telia may be preceded or accompanied by spermogonia, aecia, and uredinia or it may be the only spore form in the life cycle. Telia may vary greatly in morphology. The spores may be scattered in the mesophyll (Urediniopsis), may be within the epidermal cells (Milesina), may be grouped in one-spore-deep subepidermal crusts (Melampsora), in erumpent cushions (most species of Puccinia), or extruded as hair-like columns (Cronartium). Some telia simulate aecial types and their true nature can be determined only by germinating the spores, for example Gymnoconia (caeomatoid telia), Endocronartium (peridermioid telia), and Endophyllum and Monosporidium (aecidioid telia). These are the endocyclic "genera." Uredinoid telia (telia with Uredo morphology) are not known, but since numerous genera typically produce uredinoid aecia (Uraecium), it is probable that such a life cycle type may exist. Such a telium could be distinguished from an uredinium only by germination of the spores. Endocyclic "genera" are theoretically possible as derivatives also from Roestelia and Elateraecium. Teliospores sometimes are called "resting" or "winter" spores. This is because many common rust fungi, e.g., Puccinia graminis and Melampsora lini, have teliospores capable of surviving unfavorable periods. In fact, many teliospores will germinate only after surviving such periods. But there are rust fungi, perhaps half of all of them, whose teliospores germinate as soon as fully formed. Teliospores of species of Coleosporium, Cronartium, Chrysomyxa, Kuehneola, and Chaconia, among others, have nonresting spores. It is axiomatic that rust fungi that lack resistant teliospores compensate with some development elsewhere in the life cycle to permit survival in unfavorable periods. Teliospores are the most important spore state in generic distinctions. Much used features include whether the spores are sessile, catenulate, or pedicellate, whether unicellular or multicellular, and, if more than one-celled, whether the cells are arranged linearly, triquetrously, radially, etc. The number of germ pores per cell is important, their placement less so.

Basidiospores have received limited attention in classification, perhaps because taxonomic studies usually involve herbarium specimens with nonviable spores. But basidiospores do differ in size and shape and these differences have been used in descriptions of species (Cummins 1978; Kaneko 1981) and, with more study, may prove useful in defining genera.

SURFACE SCULPTURE OF SPORES.

The terminology for surface sculpturing of spores is not very precise. In part, this is because the patterns tend to intergrade. For example, echinulation and verrucosity become nearly indistinguishable when the "cones" and "warts" differ only slightly in the degree of sharpness or roundness (Fig. 5, C, D), or the verrucae may be rounded or block-like or have straight or annulate sides (Fig. 5, D-G). Striate verrucosity merges into interrupted or complete ridges (Fig. 5, L-M). Scanning electron microscopy doubtless will permit the introduction of more precise terms but few species have yet been studied.

The surface sculpturing of spores contributes little to generic definition but it is used extensively to delimit species. Most species of Puccinia and Uromyces on the monocots have smooth teliospores but many patterns occur in these genera. Teliospores of the Melampsoraceae, Coleosporiaceae, Cronartiaceae, Phakopsoraceae, Mikronegeriaceae, and Chaconiaceae have smooth walls. Most teliospores of the Uropyxidaceae, Pileolariaceae, and Sphaerophragmiaceae have some

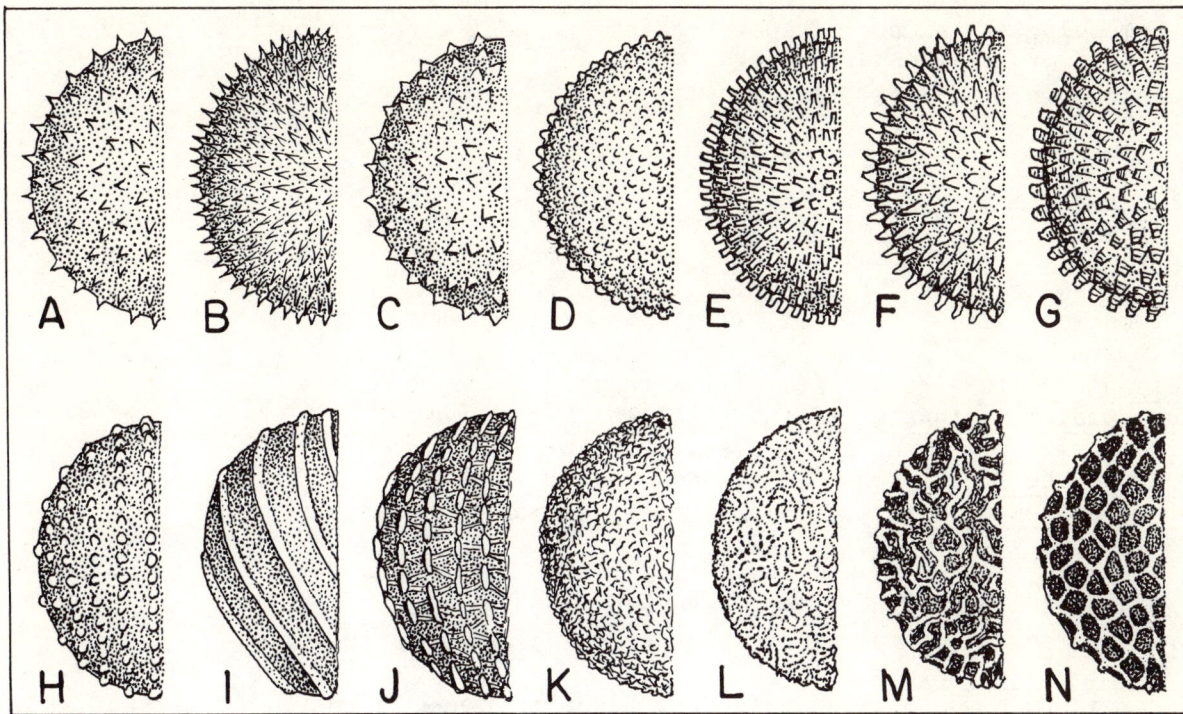

Fig. 5. Surface sculpturing of spores; not restricted as to spore state.
A-C. Echinulation; common on urediniospores; length, narrowness, and spacing
vary. D-G. Verrucose; common on aeciospores. H. Striately verrucose. I. With
ridges which may be straight, spiral, or radiating from a point and may be
beaded or crested. J. Interruptedly ridged; may be straight or spiral and may
have fine basal connections; tends to merge with striately verrucose type.
K. Rugose; roughened but with no discrete pattern. L. Labyrinthiform or cer-
ebroid. M. Pseudoreticulate; irregular and incomplete netting. N. Reticulate;
regular netting, ridges and pits.

sort of surface ornamentation. There is little consistency in the Pucciniaceae,
Phragmidiaceae, and Raveneliaceae.

Throughout the order, the majority of urediniospores are echinulate. But in
Chrysomyxa and Coleosporium the surface generally is verrucose, but the pattern of
distribution as well as the structure of the verrucae varies. Sato and Sato (1982)
have distinguished eight types of verrucosity in the aeciospores of Coleosporium
and, because the aeciospores and urediniospores of any one species tend to be
alike, it is probable that the urediniospores will also prove to have variable
surfaces. The species of Uromycladium and Atelocauda have verrucose or reticulate
urediniospores and those of Pileolaria are verrucose or ridged, usually in longitudinal or spiral patterns.

Most aeciospores that are produced in peridiate aecia (Peridermium, Aecidium,
Roestelia) have verrucose walls. But the aeciospores produced in uredinoid aecia
(Uraecium) are as the urediniospores of the species and, hence, the majority are
echinulate.

LIFE CYCLES OF THE RUST FUNGI

Depending on the number of spore states, three basic types of life cycles are
recognized and called macrocyclic, demicyclic, and microcyclic. The macrocyclic
life cycle has all spore states, the demicyclic lacks the uredinial state, and the
microcyclic cycle lacks both the aecial and the uredinial states, thus possess
only spermogonia and telia. Spermogonia may be absent from each type but especially the microcyclic life cycle.

In macrocyclic and demicyclic life cycles, the rust may be either host alternating (heteroecious), i.e., the aecial state is on one kind of plant but the telial state on a different and unrelated plant, or non-host alternating (autoecious),
i.e., the aecial and telial states on the same kind of plant.

Combining the number of spore states and host alternating features the following basic variations are recognized:
1. Heteromacrocyclic: (0), I - II, III.
   Examples:
   Coleosporium asterum (Diet.) Syd. (pine needle rust).
   Cronartium ribicola J.C. Fisch. ex Rabh. (white pine blister rust).
   Melampsorella caryophyllacearum Schroet. (fir broom rust).
   Puccinia graminis Pers. (black stem rust of wheat).
   Uromyces striatus Schroet. (alfalfa rust).
2. Automacrocyclic: (0), I, II, III.
   Examples:
   Melampsora lini (Pers.) Lév. (flax rust).
   Phragmidium mucronatum (Pers.) Schlecht. (rose rust).
   Pileolaria terebinthi (DC.) Cast. (Pistacia rust).
   Puccinia helianthi Schw. (sunflower rust).
   Uromyces appendiculatus (Pers.) Unger (common bean rust).
3. Heterodemicyclic: (0), I - III.
   Examples:
   Chrysomyxa arctostaphyli Diet. (spruce broom rust).
   Gymnosporangium juniperi-virginianae Schw. (cedar-apple rust).
   Puccinia interveniens Beth.
   Pucciniastrum goeppertianum (Kuehn) Kleb. (fir needle rust).
4. Autodemicyclic: (0), I, III.
   Example:
   Arthuriomyces peckianus (Howe) Cumm. & Y. Hirat. (orange rust of Rubus).
5. Microcyclic: (0), III.
   Examples:
   Coleosporium pinicola (Arth.) Arth. (pine needle rust).
   Puccinia malvacearum Bert. ex Mont. (hollyhock rust).
   Ravenelia pringlei Cumm. (Acacia broom rust).
   Tranzschelia thalictri (Chev.) Diet.
   Endophyllum sempervivi (Alb. & Schw.) deBary.
   Endocronartium harknessii (J.P. Moore) Y. Hirat.
   Gymnoconia nitens (Schw.) Kern & Thur.
   The last three examples represent the endocyclic life cycle discussed earlier.

THE CONCEPT OF CORRELATED SPECIES AND TRANZSCHEL'S LAW

It is generally believed that the heteroecious macrocyclic life cycle is the primitive existing one and that the shortened cycles are derivative types (Jackson 1931). Species of differing cycle but obviously close relationship often are called "correlated species." It is assumed that the species with reduced life cycles are descendants of macrocyclic species. Arthur (1934), in the Manual of the rusts in United States and Canada, made extensive use of the concept of correlated species. Puccinia interveniens Beth., P. graminella Diet. & Holw., P. sherardiana Koern., and Endophyllum tuberculatum (Ell. & Kell.) Arth. & Fromme (pages 131-133 of the Manual) represent a group of correlated species, including a heterodemicyclic, and autodemicyclic, a telioid microcyclic, and an endo or aecidioid microcyclic form. In this series the presumed macrocyclic parent is unknown but, if extant, it should produce aecia on the Malvaceae and telia on Stipa (Gramineae). In the case of Puccinia recondita Rob. ex Desm. (P. rubigo-vera Wint.) (pages 177-185 of the Manual) there is included, in addition to microcyclic species of Puccinia, also Uromyces dactylidis Otth which corresponds to the macrocyclic Puccinia in all respects except that it produces one-celled teliospores. Arthur recognized the existence of correlated species in several genera, When a rust species with a longer life cycle has been reduced to a microcyclic species there is a definite rule or pattern as to the habit of its telial state. The pattern is called Tranzschel's Law.

The essence of Tranzschel's Law is this; the telia of microcyclic species simulate the habit of the aecia of the parental macrocyclic species and occur on the aecial host plants of the latter. For example, if the aecial state of a macrocyclic species is systemic the derived microcyclic species also will have systemic mycelium. Puccinia monoica Arth. and P. thlaspeos Schub. (P. holboellii Koern) illustrate such a situation. The heteroecious P monoica, with uredinia and telia on Junegrass (Koeleria cristata (L.) Pers.), produces its systemic aecial state on species of Arabis and some other genera of the Cruciferae. P. thlaspeos, a microcyclic species, produces its systemic telia on Arabis and relatives. The teliospores of the macrocyclic and microcyclic species are morphologically similar, but the microtelia mimic the habit of the aecia of the macrocyclic parent. The microtelia, in this example, do not have a peridium but the aecial appearance is sufficiently obvious that the species was once named Aecidium holboellii by Hornemann. Another interesting pair of species that illustrates Tranzschel's Law is Puccinia coronata (crown rust), a heteroecious macrocyclic species with aecia on Rhamnus and uredinia and telia on the oat, and Puccinia mesnieriana Thuem., a microcyclic species with telia on Rhamnus. In both species the teliospores have apical digitate projections (the corona which gives crown rust both its common name and its Latin epithet). Many more examples are know, most of them involving telioid microcyclic species, but endo type microcyclic species also occur. Examples are the heteroecious demicyclic Puccinia graminella and Endophyllum tuberculatum, and Arthuriomyces peckianus and Gymnoconia nitens.

HOST-RUST RELATIONSHIP

There are definite patterns of host plant groups and the rust fungi that parasitize them. Some genera of the rust fungi, especially Puccinia and Uromyces, are comprised of species that are capable of parasitizing plants of many families. But many genera appear to be rather definitely restricted to certain plants. Host restriction may, in heteroecious species, apply to both phases of the life cycle or to only one phase. Some examples of host-rust association are discussed below. Such restrictions may be taxonomically useful because of the short-cuts to identity which they provide, but there is nothing inviolable about them. Whether or not exceptions are known, there is always the possibility that such occur. When one tries to separate genera on their own merits and without regard to the host it becomes obvious that too much emphasis has sometimes been given to the host relationship.
A. Host-restricted heteroecious rust fungi.
1. Probably the most primitive existing rust fungi are those that produce telia on ferns and aecia of firs. There are three genera that alternate between ferns and fir without known exceptions. These are Uredinopsis, Milesina, and Hyalopsora
2. There are two genera, Coleosporium and Cronartium, whose aecial states occur on pines. Coleosporium produces its aecia on the needles; Cronartium produces its aecia on stems or cones but never on needles. Neither genus is restricted as to telial hosts.
3. The genus Chrysomyxa produces its aecial state on needles or cones of spruce and the telia state on plants of the Ericaceae and close relatives.
4. Melampsora is a genus whose species are in part heteroecious and in part autoecious. The heteroecious species produce telia on willows or poplars of the Salicaceae.. Many of these produce aecia on fir, Douglas fir, or larch but others parasitize Ribes, Saxifraga, or Allium. The autoecious species are not host-restricted.
5. With few exceptions, the species of Gymnosporangium are unique in that the coniferous plants serve as the telial rather than the aecial hosts.
B. Host-restricted autoecious rust fungi.
1. Cumminsiella, a relatively small genus of macrocyclic species, appears to be restricted to species of Berberis-Mahonia.
2. Hamaspora is another small genus, whose species occur in Africa, Asia, and Australia. Its species parasitize members of the genera Rosa and Rubus, especially the latter.
3. Phragmidium, a relatively large genus, is restricted to plants of the Rosaceae. Most of the species occur on Potentilla, Rosa, and Rubus.
4. Ravenelia is one of the larger genera and is primarily tropical or semi-

tropical in distribution. The majority (perhaps all) of its species parasitize legumes.

5. The species of Prospodium occur only on the Bignoniaceae and Verbenaceae, with most of them on the Bignoniaceae.

Certain of these rust-host associations are useful in providing short-cuts to acquaintance with the commoner genera. For example, (a) if one has a rust on juniper or its relatives the rust is a Gymnosporangium, (b) if one has a rust of pine needles it is a species of Coleosporium, or (c) if the rust is on stems or cones of pine the genus is Cronartium. In temperate regions, rusts of the Rosaceae (excluding those with pomes or drupes) are likely to be species of Phragmidium but in warmer regions, Hamaspora and Gerwasia are probable. Species of Uromyces are common on legumes but one looks askance at a species of Puccinia on this family because they are so rare. They are apt to belong in Uropyxis or Sorataea. Although such short-cuts are useful, they need to be traveled with caution because exceptions exist and one should not be led so far astray as to look upon the host as a "character" of the fungus

FAMILIES OF THE UREDINALES

Two families, Melampsoraceae and Pucciniaceae, have been recognized by several authors, including Dietel (1928), Cunningham (1931), Arthur (1934), Bessey (1950), Hiratsuka (1955) and Azbukina (1974). Gaeumann (1949) accepted these two and also Pucciniastraceae, Cronartiaceae, Chrysomyxaceae, and Coleosporiaceae. Wilson and Henderson (1966) used Coleosporiaceae, Melampsoraceae, and Pucciniaceae.

Families have also been divided into subfamilies, tribes, etc. The most recent such treatment is that of Azbukina (1974) who divided the Melampsoraceae into eight subfamilies and nine tribes and the Pucciniaceae into five subfamilies and nine tribes. Most classifications emphasize teliospore morphology. Hiratsuka and Cummins (1963) questioned too extensive emphasis of the telial state. They pointed to the importance of spermogonial structure and defined 11 types. Hiratsuka and Hiratsuka (1980) added type 12 and, based on a study of 224 species in 73 genera, tabulated the genera in six groups based on spermogonial morphology. With emphasis on spermogonial type we propose the following 14 families as associating most of the genera in relatively natural groups that probably reflect relationship better than previous systems. We have not constructed the keys using families but the sections approximate the families.

PUCCINIASTRACEAE (Arth.) Gaeumann.

Spermogonia type 1, 2, or 3; group I. Aecial state Peridermium with cylindrical peridium, in two species Uraecium, aeciospores catenulate, except in Uraecium, mostly verrucose. Uredinia with cellular peridium, opening by a pore, often with differentiated ostiolar cells, spores borne singly, mostly echinulate, pores obscure, scattered or bizonate. Telia either subepidermal or intradermal, not erumpent, composed of laterally adherent spores one spore deep, spores multicellular by vertical septa or unicellular, sessile, germ pore obscure, mostly one, perhaps not always differentiated, basidia external. Species heteroecious and mostly macrocyclic with aecia on conifers.

TYPE: Pucciniastrum Otth.
Genera: Hyalopsora, Melampsorella, Melampsoridium, Milesina, Pucciniastrum, Uredinopsis.

COLEOSPORIACEAE Dietel.

Spermogonia type 2; group I. Aecial state Peridermium, the peridium mostly strongly developed, aeciospores catenulate and verrucose. Uredinia with rudimentary peridium or none, spores catenulate and mostly verrucose, similar to the aeciospores of the species, pores mostly obscure, scattered. Telia erumpent, waxy or gelatinous, hard when dry, pulvinate or columnar, teliospores unicellular, sessile, catenulate, pseudocatenulate, or in a singly layer, walls thin, pores not differentiated, germinating without dormancy either by septation of the spore (internal basidium) and the production of sterigmata (Coleosporium), or by an external basidium with short sterigmata (Chrysomyxa). Most species heteroecious with the aecial state on needles or cones of conifers.

TYPE: Coleosporium Léveillé.

Genera: Chrysomyxa, Coleosporium.

CRONARTIACEAE Dietel.

Spermogonia type 9; group II. Aecial state Peridermium, the peridium large and blister-like, strongly developed, of one or more layers of cells, rupturing widely, aeciospores catenulate, coarsely verrucose with rods or platelets often with annulate sides. Uredinia with peridium and differentiated ostiolar cells, sometimes also with intrasoral paraphyses, urediniospores borne singly, echinulate, pores bizonate. Telia columnar, firm, with peridium as the uredinia, teliospores unicellular, embedded in a common matrix, pores one to three, germinating without dormancy, basidium external. Species (except the microcyclic) heteroecious and macrocyclic with aecial state on stems and cones of pines.

TYPE: Cronartium Fries.

Genera: Cronartium, Endocronartium (endocyclic).

MICRONEGERIACEAE Cumm. & Y. Hirat. fam. nov.

Spermogoniis profunde immersis, typus 12. Aeciis Caeoma, sporis catenulatis, verrucosis. Urediniis sine peridio vel paraphysibus, sporis singulatim natis, echinulatis, poro germinationis non visis forte nullis. Teliis erumpentibus, ceraceis, sporis sessilis, singulis, unicellularibus, lateraliter discretus, pariete uniformiter tenuis, poro germinationis nullis, statim germinantibus.

Spermogonia type 12; group III. Aecial state Caeoma, without precise peridium but with hyphoid cellular layers over spore mass, aeciospores catenulate, verrucose. Uredinia without peridia or paraphyses, urediniospores borne singly, echinulate, pores not seen, perhaps not differentiated. Telia waxy in appearance, without peridium or paraphyses, teliospores sessile, unicellular, laterally free, thin-walled, germinating without dormancy by continuing growth apically, development of septa, sterigmata, and basidiospores, the entire protoplast used in the process, the basal structure then collapsing. Species heteroecious with the aecial state on conifers and the telial state on Fagaceae.

TYPE: Mikronegeria Dietel.

Genera: Mikronegeria.

MELAMPSORAECEAE Schroeter.

Spermogonia type 2 or 3; group I. Aecial state Caeoma, with rudimentary or no peridium and catenulate, verrucose spores. Uredinia with abundant capitate paraphyses and sometimes also rudimentary peridium, urediniospores borne singly, echinulate, pores scattered or bizonate. Telia subepidermal or rarely subcuticular, not erumpent, consisting of laterally adherent spores in crusts one spore deep or some species also with subjacent spore-like (sterile?) cells, spores unicellular, sessile, pigmented, pore one, basidium external. Species heteroecious or autoecious; mostly macrocyclic.

TYPE: Melampsora Castagne.

Genera: Melampsora.

PHAKOPSORACEAE (Arth.) Cumm. & Y. Hirat. stat. nov.

Phakopsoreae Arth., Manual of the rusts in United States and Canada. p. 1. 1934.

Spermogonia type 5, 7; group VI. Aecial state Aecidium, Caeoma, or Uraecium, thus without or with peridium, aeciospores catenulate or borne singly, verrucose or echinulate. Uredinia mostly with basally united, peripheral, incurved, dorsally thick-walled paraphyses, urediniospores borne singly except in Arthuria where catenulate, echinulate, pores obscure, mostly scattered. Telia erumpent or embedded in host tissue, two to several spores deep, teliospores unicellular, sessile, catenulate or irregularly arranged, germ pore one in each cell or perhaps not differentiated in some, basidium external. Species heteroecious or autoecious; not host restricted.

TYPE: Phakopsora Dietel.

Genera: Arthuria, Cerotelium, Crossopsora, Dasturella, Nothoravenelia, Phakopsora, Phragmidiella, Physopella, Pucciniostele, Uredopeltis, Monosporidium (endocyclic).

CHACONIACEAE Cumm. & Y. Hirat. fam. nov.

Spermogoniis typus 5 vel 7. Aeciis Uraecium vel Aecidium ergo sporis singulis vel catenulatis, echinulatis vel verrucosis. Urediniis cum vel absque paraphysibus, sporis singulis, echinulatis, poris variis. Teliis erumpentibus, sporis singulis, unicellularis lateraliter dicretus, sessiles vel pedicellatis, pariete uniformiter tenuis, plerumque hyalinis, poro germinationis unus vel nullis, statim germinantibus.

Spermogonia type 5, 7. Aecial state Uraecium or Aecidium, thus with or without peridium, aeciospores borne singly or catenulately, mostly echinulate, pores various. Uredinia with or without paraphyses, spores borne singly, echinulate for most, pores various. Telia erumpent, teliospores unicellular, laterally free, sessile or pedicellate, thin-walled, with or without a poorly defined germ pore, germinating without dormancy, basidium external, or internal by formation of septa within the spore. Species autoecious or heteroecious; hosts various.

TYPE: Chaconia Juel.

Genera: Achrotelium, Aplopsora, Botryorhiza, Ceropsora, Chaconia, Chrysocelis Goplana, Maravalia, Ochropsora, Olivea. Note: Chrysocelis has type 4 spermogonia but otherwise should belong here.

UROPYXIDACEAE (Arth.) Cumm. & Y. Hirat. stat. nov.

Uropyxioideae Arth. Reś. Sci. Congr. Internat. Bot. Vienne. p. 341. 1906.

Spermogonia type 5, 7; group VI. Aecial state Aecidium, Caeoma, or more commonly Uraecium, thus with or without peridium, or with paraphyses, aeciospores catenulate or borne singly, verrucose or echinulate. Uredinia with or without paraphyses, urediniospores borne singly, mostly echinulate, pores mostly scattered. Telia with or without paraphyses, teliospores two- or more-celled by horizontal septa, often with bilaminate wall the outer layer hygroscopic, pore one or more per cell, basidium external, spores pedicellate, the pedicels often hygroscopic. Species autoecious or heteroecious; hosts various.

TYPE: Uropyxis Schroeter.

Genera: Dasyspora, Didymopsorella, Dipyxis, Macruropyxis, Newinia, Phragmopyxis, Porotenus, Prospodium, Sorataea, Tranzschelia, Uropyxis.

PILEOLARIACEAE (Arth.) Cumm. & Y. Hirat. stat. nov.

Pileolarieae Arth. Rés. Sci. Congr. Internat. Bot. Vienne. p. 340. 1906.

Spermogonia type 5, 7; group VI. Aecial state Uraecium, spores borne singly, usually reticulate, ridged, verrucose, or spirally marked, pores zonate. Uredinia as the aecia but unaccompanied by spermogonia. Telia erumpent, teliospores unicellular with one or a few spores on each pedicel which may also have sterile cyst-like cells laterally, germinating by external basidium, pore one per spore. Species autoecious; hosts Anacardiaceae and Leguminosae.

TYPE: Pileolaria Castagne.

Genera: Atelocauda, Pileolaria, Uromycladium.

RAVENELIACEAE (Arth.) Leppik.

Spermogonia type 5, 7; group VI. Aecial state Aecidium with peridium or typically Uraecium with or without paraphyses and spores borne singly, echinulate in most. Uredinia with or without paraphyses, spores borne singly, mostly echinulate, pores various, usually obvious. Telia erumpent, with or without paraphyses, spores pedicellate, vertically septate or vertically or radially arranged atop the pedicel, of two or more spores or of two to many cells, often subtended by hygroscopic cysts or with pedicels having apical cells, basidium external, pores one or two in each cell or spore. Species autoecious; hosts mostly Leguminosae.

TYPE: Ravenelia Berkeley.

Genera: Anthomyces, Anthomycetella, Apra, Cystomyces, Diabole, Dicheirinia, Diorchidiella, Diorchidium, Kernkampella, Lipocystis, Ravenelia, Sphenospora, Spumula, Ypsilospora.

PHRAGMIDIACEAE Corda.

Spermogonia type 6, 8, 10, 11; group IV. Aecial state Caeoma, with catenulate spores, or Uraecium, with spores borne singly, aeciospores verrucose or echinulate. Uredinia mostly with thin-walled, incurved, peripheral paraphyses, urediniospores borne singly, mostly echinulate, pores scattered. Telia erumpent, with or without paraphyses, spores with one to several cells by horizontal septa, pedicellate, pores one or more in each cell, basidium external. Species autoecious, mostly and perhaps only on Rosaceae.

TYPE: Phragmidium Link.

Genera: Arthuriomyces, Frommeëlla, Gerwasia, Hamaspora, Joerstadia, Kuehneola, Phragmidium, Trachyspora, Xenodochus, Gymnoconia (endocyclic).

SPHAEROPHRAGMIACEAE Cumm. & Y. Hirat. fam. nov.

Spermogoniis subepidermalibus, typus 5, vel subcuticularibus, typus 7. Aeciis Uraecium vel Aecidium ergo sporis singulis vel catenulatis. Urediniis plerumque cum paraphysibus peripherales, sporis singulis, echinulatis, poris germinationis varius. Teliis erumpentibus, sporis pedicellatis, 3-cellularibus, triquetrus vel cum cellulis quatuor vel plus quam quatuor et murifomiter septatis, poris germinationis unus vel alter.

Spermogonia types 5, 7, 11; groups IV, VI. Aecial state Uraecium, with echinulate spores borne singly, or Aecidium, with catenulate, verrucose spores. Uredinia mostly with peripheral paraphyses, spores borne singly, echinulate, germ pores equatorial or scattered. Telia erumpent, with or mostly without paraphyses, spores pedicellate, triquetrously arranged or with four or more spores in muriformly septate heads, germ pore one or two in each spore or cell, basidium external. Species autoecious; mostly on Leguminosae or Rosaceae.

TYPE: Sphaerophragmium Magnus.

Genera: Cumminsina, Hapalophragmium, Nyssopsora, Sphaerophragmium, Triphragmiopsis, Triphragmium. Note: Triphragmium has type 11 spermogonia but otherwise belongs here.

PUCCINIACEAE Chevalier.

Spermogonia type 4; group V. Aecial state Aecidium, with peridium and catenulte, mostly verrucose spores, or Uraecium, without peridium and with spores borne singly, or rarely Caeoma, without peridium but with catenulate spores. Uredinia with or without paraphyses or rarely with palisade-like peridium, spores borne singly, mostly echinulate, pores various. Telia with or without paraphyses but rarely with palisade-like peridium, or telia may be separated into locules by stromatoid paraphyses, teliospores borne singly, mostly pedicellate, with one or two (rarely more) cells, septa horizontal or oblique, germ pore mostly one in each cell, germination mostly by external basidium, rarely by internal basidium. Species heteroecious or autoecious; hosts various.

TYPE: Puccinia Persoon.

Genera: Chrysella, Chrysocyclus, Chrysopsora, Cleptomyces, Corbulopsora, Cumminsiella, Gymnosporangium, Kernella, Miyagia, Polioma, Puccinia, Stereostratum, Uromyces, Zaghouania, Endophyllum (endocyclic).

PUCCINIOSIRACEAE (Diet.) Cumm. & Y. Hirat. stat. nov.

Pucciniosireae Diet. Engl. & Prantl Nat. Pflanzenfam. 6:93. 1928.

Spermogonia type 4; group V. Aecial and uredinial state not produced, all genera microcyclic. Telia mostly resembling in gross appearance aecial states of macrocyclic genera, with or without peridium, teliospores sessile, of one or two cells, often catenulate with intercallary cells, or extruded as columns or filaments, the columns often firm, basidia external or internal. Species all autoecious and presumably derivatives of macrocyclic genera but the relationship often is obscure. This is by nature a heterogeneous family.

TYPE: Pucciniosira Lagerheim.

Genera: Alveolaria, Baeodromus, Chardoniella, Cionothrix, Didymopsora, Dietelia, Pucciniosira, Trichopsora, unnamed genus.

GENERA OF UNCERTAIN AFFINITIES: Blastospora, Desmella, Edythea, Hemileia, Hiratsukamyces, Kweilingia, Masseeëlla, Skierka.

EXCLUDED GENUS: Hiratsukaia.

COLLECTION, PRESERVATION, AND IDENTIFICATION OF RUST FUNGI

Proper procedures for collecting and preserving specimens are important. It is essential to collect enough plant parts, such as flowers, fruits and uninfected leaves, together with the infected parts, so that the host may be identified. this is especially important for such plants as sedges and grasses. Without the inflorescence these plants are very nearly impossible to identify.

A good hand lens is recommended for use in the field to check the presence of different spore states or to distinguish insects or other pathogens than the rust fungi.

Because all spore states can be preserved dry, specimens can be handled just as flowering plant specimens, using plant presses. Dried and pressed specimens usually are kept in specimen packets bearing labels with the essential data, such as the name of the host, name of the collector, the date and location where found, and preferable a collection number. Packet specimens may be attached to standard herbarium sheets and stored as are flowering plants. Some mycologists prefer to use a standardized 5 x 7 in packet and store in boxes or filing cases. Either system works; it is a matter of personal choice.

It is essential for the identification of these fungi to have a compound microscope with a magnification of at least 400x and equiped with a calibrated micrometer. Spores are best mounted in lactophenol or some similar non- or slow-drying medium with some clearing action. Stains are generally not needed. Size, surface ornamentation, shape, and color should be recorded reasonably promptly. If stored for a long time, some spore walls change considerably, as does size. Thin sections made free hand or even with a microtome may be useful or necessary to determine the position of a sorus in the host tissues or to observe the structure of the fungus.

If a comprehensive regional descriptive manual is available, the host index may be used to narrow the possibilities and then each likely species can be compared with the specimen being identified. With such a manual, one may key out the genus and then refer to the literature cited to determine specific identity. In some cases, identification may be confirmed by comparing with authentically identified specimens or by sending to some specialist.

LITERATURE CITED

Arthur, J.C. 1905. Terminology of the spore-structures in the Uredinales. Bot. Gaz. 39:219-222.

_____ . 1925. Terminology of the Uredinales. Bot. Gaz. 80:219-223.

_____ . 1929. The Plant Rusts. Wiley & Sons. New York. 446 pp.

_____ . 1934. Manual of the Rusts in United States and Canada. Purdue Res. Found. Lafayette, IN. 438 pp.

Azbukina, Z.M. 1974. Rust Fungi of the Soviet Far East (in Russian). Akad. Nauk CCCP. Moscow. 527 pp.

Bessey, E.A. 1950. Morphology and Taxonomy of Fungi. Blakiston, Philadelphia. 791 pp.

Charudattan, R., McKinney, D.E., Cordo, H.A., and Silveria-Guideo, H.A. 1976. Uredo Eichhorniae, a potential biocontrol agent for water hyacinth. In Proc. 4th. Int. Sympos. Biol. Control of Weeds (ed. T.F. Freeman). p. 210-213.

Cummins, G.B. 1936. Phylogenetic significance of the pores in urediospores. Mycologia 28:103-132.

_____ . 1959. Illustrated Genera of Rust Fungi. Burgess. Minneapolis. 131 pp.

_____ . 1978. Rust Fungi on Legumes and Composites in North America. Univ. Ariz. Press. Tucson. 424 pp.

Cunningham, G.H. 1930. Terminology of the spore forms and associated structures of the rust fungi. N. Zeal. J. Sci. Tech. 12:123-128.

Dietel, P. 1928. Reihe Uredinales. In Engler & Prantl Nat. Pflanzenfam. 6:24-97.

Durrieu, G. 1979. Les cycles des Uredinales, Problems de terminologie. Bull. Soc. Mycol. Fr. 95:379-392.

Gäumann, E. 1949. Die Pilze. Birkhauser, Basil. 382 pp.

Hasan, S. and Walshere, A.J. 1973. The biology of Puccinia chondrillina, a potential biological control of skeleton weed. Ann. Appl. Biol. 74:325-332.

Hiratsuka, N. 1955. Uredinological Studies (in Japanese). Kasai Publ. and Printing CO. Tokyo. 382 pp.

Hiratsuka, Y. 1973a. Sorus development, spore morphology and nuclear condition of Gymnosporangium gaeumannii ssp. albertensis. Mycologia 65:137-144.

_____. 1973b. The nuclear cycle and terminology of spore states in Uredinales. Mycologia 65:432-443.

_____. 1975. Recent controversies and the terminology of the rust fungi. Rept. Tottori Mycol. Inst. 12:99-104.

_____, and Cummins, G.B. 1963. Morphology of spermogonia of the rust fungi. Mycologia 55:487-507.

_____, and Hiratsuka, N. 1980. Morphology of spermogonia and taxonomy of rust fungi. Rept. Tottori Mycol. Inst. 18:257-268.

Holm, L. 1973. Some notes on rust terminology. Rept. Tottori Mycol. Inst. 10:183-187.

Jackson, H.S. 1931. Present evolutionary tendencies and the origin of life cycles in the Uredinales. Mem. Torrey Bot. Club 18:1-108.

Kaneko, S. 1981. The species of Coleosporium, the causes of pine needle rusts, in the Japanese Archepelago. Rept. Tottori Mycol. Inst. 19:1-159.

Kenney, M.J. 1970. Comparative morphology of the uredia of the rust fungi. Ph.D. thesis. Purdue University. 87 pp.

Laundon, G.F. 1965. The generic names of Uredinales. Commonw. Mycol. Inst. Mycol. Papers. No. 102:1-24.

_____. 1967. Terminology in the rust fungi. Trans. Brit. Mycol. Soc. 50:189-194.

_____. 1967. Delineation of aecial from uredial states. Trans. Brit. Mycol. Soc. 344-346.

Oehrens, E. 1977. Biological control of the blackberry through introduction of rust. FAO Plant Prot. Bull. 25:26-28.

Sathe, A.V. 1977. Morphology and classification of uredinia. Kavaka 5:59-64.

Sato, T. and Sato, S. 1982. Aeciospore surface structure of the Uredinales. Trans. Mycol. Soc. Japan 25:51-63.

Savile, D.B.O. 1973. Aeciospore types in Puccinia and Uromyces attacking Cyperaceae, Juncaceae and Poaceae. Rept. Tottori Mycol. Inst. 10:225-241.

Thirumalachar, M.J. and Mundkur, B.B. 1949, 1950. Genera of rusts. I, II, III. Indian Phytopathol. 2:65-101; 193-244; 3:4-42.

Wilson, M. and Henderson, D.M. 1966. British Rust Fungi. Cambridge Univ. Press. 344 pp.

KEY TO THE GENERA

KEY TO THE SECTIONS

1. Teliospores sessile (may be in chains) 2
1. Teliospores pedicellate 8
  2. Telia 1 spore deep 3
  2. Telia 2 or more spores deep 4
3. Telia covered by host, not erumpent (genera 1-8) .... Section I
3. Telia rupturing host, erumpent (genera 12-20) .... Section III
  4. Telia covered by host, not erumpent (genera 9-11) .... Section II
  4. Telia rupturing host, erumpent 5
5. Telia low or only cushion-like 6
5. Telia columnar or filiform 7
  6. Teliospores 1-celled (genera 12, 21-29, 37) .... Section IV
  6. Teliospores 2-celled (genus 38) .... Section VI
7. Teliospores 1-celled (genera 12, 21, 29, 30-36, 37) .... Section V
7. Teliospores 2-celled (genera 39, 40) .... Section VII
  8. Teliospores 1-celled (genera 41-55, 77) .... Section VIII
  8. Teliospores 2- or more-celled 9
9. Teliospores horizontally (rarely vertically) septate 10
9. Teliospore septation otherwise 11
  10. Teliospores 2-celled (genera 56-75, 78, 79, 85) .... Section IX
  10. Teliospores 3- or more-celled (genera 59, 63, 67, 76, 80-85) Section X
11. Teliospores 1-celled in free binate pairs or 2-4-celled by
    vertical septa; pedicels mono-hyphal, with or without apical
    cells (genera 57, 58, 70, 86-93) .... Section XI
11. Teliospores and/or septation otherwise 12
  12. Teliospores laterally united in radial heads with multi-
     hyphal pedicels or if mono-hyphal then with hygroscopic cysts
     (genera 94-99) .... Section XII
  12. Teliospores triquetrously or muriformly septate (genera
     100-105) .... Section XIII

ADDENDUM: endocyclic "genera." These genera resemble the various kinds of aecia
as to sorus and spores but the spores germinate by basidia and can be distinguished only by spore germination. Thus such "genera" are actually only
microcyclic life cycle variants in the same sense as "micropuccinia" but
derived from the aecial rather than the telial state.

1. As _Aecidium_; spermogonia subepidermal, type 4 .... Endophyllum 63E
2. As _Aecidium_; spermogonia subcuticular, type 7 .... Monosporidium 9E
3. As _Caeoma_ .... Gymnoconia 78E
4. As _Peridermium_ .... Endocronartium 30E
5. As _Elateraecium_, theoretically possible but unknown.
6. As _Roestelia_, theoretically possible but unknown.
7. As _Uraecium_, theoretically possible but unknown.

KEY TO GENERA OF SECTION I

1. Teliospore wall colorless 2
1. Teliospore wall yellowish brown to brown 6
  2. Urediniospores white, devoid of pigment   3
  2. Urediniospores with yellow to orange protoplasmic
       pigment, wall colorless 4
3. Teliospores in the mesophyll .... Uredinopsis (1)
3. Teliospores within epidermal cells .... Milesina (2)
  4. Teliospores subepidermal, ostiolar cells of uredinia
      elongate, pointed .... Melampsoridium (5)
  4. Teliospores intraepidermal, ostiolar cells not elongate 5
5. Uredinial peridium poorly developed, opening irregularly .. Hyalopsora (3)
5. Uredinial peridium well-developed, ostiole discrete .... Melampsorella (4)
  6. Uredinia with well-developed peridium and ostiolar cells Pucciniastrum (6)
  6. Uredinia with rudimentary or no peridium 7
7. Uredinia with capitate paraphyses .... Melampsora (7)
7. Uredinia bounded by fused hyphae and host cells .... Hiratsukamyces (8)

KEY TO GENERA OF SECTION II

1. Teliospores not catenulate, irregularly arranged .... Phakopsora (9)
1. Teliospores catenulate 2
  2. Aecia Aecidium; telia usually 2-4 spores deep ...... Physopella (10)
  2. Aecia Caeoma; telia 6-10 spores deep .... Pucciniostele (11)

KEY TO GENERA OF SECTION III

1. Teliospores becoming basidia by septation 2
1. Teliospores producing an external basidium 4
  2. Telia not gelatinous .... Ochropsora (14)
  2. Telia gelatinous when wet 3
3. Urediniospores catenulate, verrucose .... Coleosporium (12)
3. Urediniospores borne singly, echinulate .... Goplana (13)
  4. Telia waxy-gelatinous when moist 5
  4. Telia otherwise 7
5. Only telia known, probably microcyclic .... Ceropsora (16)
5. Telia, aecia, uredinia known, macrocyclic 6
  6. Spermogonia type 12; basidium very broad .... Mikronegeria (15)
  6. Spermogonia type 4; basidium narrow .... Chrysocelis (17)
7. Urediniospores radially asymmetrical, lobed .... Olivea (20)
7. Urediniospores radially symmetrical, not lobed 8
  8. Teliospores grouped on basal sporogenous cells .... Chaconia (19)
  8. Teliospores without such basal cells .... Aplopsora (18)

KEY TO GENERA OF SECTION IV

1. Species producing only telia; spermogonia type 4 2
1. Species producing uredinia and telia; spermogonia where
    known not type 4 3
  2. Telia composed of radial plates of spores .... Alveolaria (28)
  2. Teliospores not in radial plates .... Baeodromus (29)
3. Urediniospores catenulate 4
3. Urediniospores borne singly 5
  4. Urediniospores verrucose; species heteroecious .... Chrysomyxa (21)
  4. Urediniospores echinulate; species autoecious .... Arthuria (22)
5. Teliospores spreading from the sorus as a mantle .... Kweilingia (25)
5. Teliospores remaining in discrete sori 6
  6. Teliospores only loosely adherent, pale 7
  6. Teliospores strongly adherent, brown 8

(continued next page)

7. Uredinia with peridium or peripheral paraphyses .... Cerotelium (23)
7. Uredinia without such structures .... Phragmidiella (24)
   8. Teliospores irregularly arranged .... Uredopeltis (26)
   8. Teliospores catenulate .... Dasturella (27)

KEY TO GENERA OF SECTION V

1. Species producing telia and uredinia 2
1. Species producing telia only 5
   2. Uredinia with peridium .... Cronartium (30)
   2. Uredinia without peridium 3
3. Uredinia with peripheral paraphyses .... Crossopsora (31)
3. Uredinia without paraphyses 4
   4. Teliospores embedded in a gelatinous matrix .... Masseeella (33)
   4. Teliospores without such a matrix .... Skierka (32)
5. Teliospores adherent in radial plates .... Alveolaria (28)
5. Teliospores not in radial plates 6
   6. Teliospores becoming a basidium by septation 7
   6. Teliospores producing an external basidium 8
7. Teliospores with elongate gelatinizing intercallary cells Trichopsora (34)
7. Teliospores without such intercallary cells .... Coleosporium (12)
   8. Teliospores irregularly arranged .... Cionothrix (35)
   8. Teliospores catenulate 9
9. Telia with Aecidium-like peridium .... Dietelia (37)
9. Telia without such a peridium 10
   10. Teliospores separated by obvious intercallary cells . un-named (36)
   10. Teliospores without intercallary cells .... Chrysomyxa (21)

KEY TO GENERA OF SECTION VI

1. Teliospores Puccinia-like but sessile .... Polioma (38)

KEY TO GENERA OF SECTION VII

1. Telia with obvious intercallar cells .... Pucciniosira (39)
1. Telia without obvious intercallary cells .... Didymopsora (40)

KEY TO GENERA OF SECTION VIII

1. Telia columnar, spores separated by pedicel-like inter-
    callar cells .... Chardoniella (41)
1. Telia not columnar, pustular 2
   2. Teliospores becoming basidia by septation 3
   2. Teliospores producing external basidia 4
3. Spermogonia type 4; telia gelatinous .... Chrysella (42)
3. Spermogonia type 7; telia not gelatinous .... Achrotelium (43)
   4. Urediniospores plano-convex, the flat side smooth, con-
    vex side aculeate .... Hemileia (44)
   4. Urediniospores otherwise 5
5. Spermogonia intraepidermal, type 6 .... Gerwasia (45)
5. Spermogonia otherwise 6
   6. Basidiospores sessile, not ejected .... Zaghouania (47)
   6. Basidiospores on sterigmata, ejected 7
7. Telia suprastomatal.. Blastospora (46)
7. Telia pustular, rupturing epidermis 8
   8. Teliospores with thin, pale or colorless walls 9
   8. Teliospores with firm, usually brown or brownish walls 10
9. Teliospores small, egg-shaped .... Botryorhiza (48)
9. Teliospores elongate-ellipsoid to cylindrical .... Maravalia (49)
   10. Spermogonia type 4 11
   10. Spermogonia otherwise 12

(continued next page)

11. Uredinia and telia with palisade-like peridium .... Corbulopsora (54)
11. Uredinia and telia without peridium .... Uromyces (55)
    12. Spermogonia type 10 or 11 13
    12. Spermogonia type 5 or 7 14
13. Teliospore pedicel thin-walled, fragile .... Trachyspora (76)
13. Teliospore pedicel thick-walled, persistent .... Phragmidium (85)
    14. Teliospores borne 1-3 on branched pedicels; uredinio-
        spores verrucose or reticulate .... Uromycladium (52)
    14. Teliospores borne singly on unbranched pedicels 15
15. Teliospores on short broad pedicels; uredinia with capi-
    tate paraphyses .... Lipocystis (53)
15. Teliospores on narrow, elongate pedicels; uredinia with-
    out capitate paraphyses 16
    16. Teliospores depressed-globoid, verrucose or reticulate ... Pileolaria (50)
    16. Teliospores broadly ellipsoid or longer, smooth or
        with projections or block-like verrucae .... Atelocauda (51)

KEY TO GENERA OF SECTION IX

1. Teliospore cells becoming basidia by septation .... Chrysopsora (56)
1. Teliospores producing external basidia 2
    2. Telia suprastomatal 3
    2. Telia otherwise 5
3. Sporogenous cells emerging through stomata .... Desmella (57)
3. Sporogenous cells differentiated above stomata 4
    4. Sorus without a basal peridial cup .... Edythea (58)
    4. Sorus with a basal peridial cup .... Prospodium (70)
5. Telia conspicuously gelatinous when wet 6
5. Telia otherwise 7
    6. Teliospores horizontally septate .... Gymnosporangium (59)
    6. Teliospores vertically septate .... Sphenospora (86)
7. Teliospores extruded in hair-like columns 8
7. Teliospores not so extruded 9
    8. Telia without intermixed paraphyses .... Kernella (60)
    8. Telia with intermixed, hygroscopic paraphyses .... Didymopsorella (61)
9. Uredinia and telia with palisade-like peridium .... Miyagia (62)
9. Uredinia and telia otherwise 10
    10. Teliospore cells with 1 germ pore 11
    10. Teliospore cells with pores otherwise 19
11. Spermogonia type 8 .... Joerstadia (79)
11. Spermogonia otherwise 12
    12. Spermogonia type 4 13
    12 Spermogonia otherwise 14
13. Basidium narrow, strictly external .... Puccinia (63)
13. Basidium broad, semi-internal or giving that impression ... Chrysocyclus (65)
    14. Teliospores beset with rod-like papillae mostly
        branched at apex; spermogonia type 5 .... Dasyspora (66)
    14. Teliospore surface otherwise 15
15. Spermogonia type 6; aecia Caeoma .... Arthuriomyces (78)
15. Spermogonia otherwise; aecia Aecidium or Uraecium 16
    16. Teliospore wall smooth, hyaline or pale brownish; germ
        pore poorly or not differentiated 17
    16. Teliospore wall mostly sculptured; germ pore difer-
        entiated 18
17. Basidium produced through germ pore or discrete area
    in spore wall .... Porotenus (68)
17. Basidum produced by elongation of apex of cells .... Sorataea (67)

(continued next page)

18. Aecia _Aecidium_; uredinia with intermixed paraphyses; teliospores tending to adhere in bunches .... Tranzschelia (69)
18. Aecia _Uraecium_; uredinia with peripheral paraphyses; teliospores not adhering .... Prospodium (70)
19. Teliospores with 2 pores per cell 20
19. teliospores with more than 2 pores 21
   20. Teliospore pedicels not hygroscopic or fragile; urediniospore pores zonate; spermogonia type 4 .... Cumminsiella (71)
   20. Teliospore pedicels hygroscopic or fragile; urediniospore pores scattered, obscure; spermogonia type 7 .... Uropyxis (72)
21. Teliospores with 4 or more pores; spermogonia type 4 .... Cleptomyces (73)
21. Teliospores with 3 pores per cell 22
   22. Teliospore wall simple, smooth; spermogonia unknown Stereostratum (64)
   22. Teliospore wall tending bilaminate; spermogonia type 7, or 10 or 11 23
23. Spermogonia type 10 or 11; aecia _Caeoma_ .... Phragmidium (85)
23. Spermogonia type 7 24
   24. Teliospore germ pores appearing as if perforated .... Macruropyxis (74)
   24. Teliospore pores not appearing perforated .... Dipyxis (75)

KEY TO GENERA OF SECTION X

1. Telia conspicuously gelatinous when wet .... Gymnosporangium (59)
1. Telia otherwise 2
   2. Basidia produced by elongation of apex of cells .... Sorataea (67)
   2. Basidia otherwise 3
3. Spermogonia type 7; teliospore wall conspicuously bilaminate .... Phragmopyxis (76)
3. Spermogonia otherwise; teliospore wall not or inconspicuously bilaminate 4
   4. Spermogonia type 4; teliospores typically 2-celled but with exceptions .... Puccinia (63)
   4. Spermogonia otherwise; teliospores typically 3-or more-celled 5
5. Teliospore wall colorless 6
5. Teliospore wall brown 7
   6. Teliospores terete, tending lanceolate, in felty extrusions .... Hamaspora (80)
   6. Teliospores cylindrical, appearing articulated, not felt-like .... Kuehneola (81)
7. Spermogonia type 7; aeciospores and urediniospore walls bilaminate .... Newinia (82)
7. Spermogonia otherwise; aeciospore and urediniospore wall simple 8
   8. All teliospore cells with 1 germ pore .... Frommeëlla (83)
   8. Most teliospore cells with 2 or more germ pores 9
9. Teliospores with 2 or more pores except the apical cell .. Xenodochus (84)
9. Teliospores with 2 or more pores in all cells .... Phragmidium (85)

KEY TO GENERA OF SECTION XI

1. Teliospores 1-celled, laterally free on the pedicel 2
1. Teliospores 2- or more-celled with vertical septa 4
   2. Pedicel without apical cells .... Ypsilospora (86)
   2. Pedicel with apical cells 3
3. Each apical cell with a single spore .... Apra (87)
3. Each apical cell with a pair of spores .... Diabole (88)

(continued next page)

```
4.  Pedicel with apical cells .........................   5
4.  Pedicel without apical cells .....................   7
5.  Teliospores with 2 germ pores per cell ...............   Diorchidiella (89)
5.  Teliospores with 1 germ pore per cell ...............   6
   6.  Teliospores with coarse, block-like sculpture .......   Dicheirinia (90)
   6.  Teliospores with smooth walls .....................   Anthomyces (91)
7.  Teliospores in a gelatinous-oily matrix ..............   Sphenospora (92)
7.  Teliospores without such a matrix ...................   8
   8.  Telia suprastomatal ..............................   9
   8.  Telia pustular, rupturing epidermis ..............   Diorchidium (93)
9.  Sporogenous cells emerging through stomata ...........   Desmella (57)
9.  Sporogenous cells differentiated above stomata ........   10
   10.  Sorus with a basal peridial cup ..................   Prospodium (70)
   10.  Sorus without a peridial cup .....................   Edythea (58)
```

KEY TO GENERA OF SECTION XII

```
1.  Teliospores sessile with subtending cystoid cells; para-
      physes peripheral, incurved, basally united ..........   Nothoravenelia (95)
1.  Teliospores pedicellate ............................   2
   2.  Teliospores without cysts; paraphyses peripheral,
         long, incurved, septate basally ..................   Anthomycetella (94)
   2.  Teliospores with subtending, hygroscopic cysts ......   3
3.  Pedicel simple, mono-hyphal .........................   4
3.  Pedicel compound, multi-hyphal ......................   5
   4.  Pedicel attached to the cysts ....................   Cystomyces (96)
   4.  Pedicel attached to the spore head ...............   Spumula (97)
5.  Teliospore heads with a patelliform cellular layer
      between spores and cysts ...........................   Kernkampella (98)
5.  Teliospores heads with cysts attached to spores ........   Ravenelia (99)
```

KEY TO GENERA OF SECTION XIII

```
1.  Teliospores triquetrously septate ...................   2
1.  Teliospores muriformly septate ......................   5
   2.  Two teliospores basal surmounted by a third ........   Hapalophragmium (100)
   2.  One teliospore basal surmounted by two other ........   3
3.  Teliospores with 1 germ pore per cell ...............   Triphragmium (101)
3.  Teliospores with 2 germ pores per cell ..............   4
   4.  Teliospores verrucose with irregular warts, germ pores
         in outer walls .................................   Triphragmiopsis (102)
   4.  Teliospores with elongate, often apically divided
         projections, germ pores at inner angles ..........   Nyssopsora (103)
5.  Cells of teliospore in a more or less globoid arrange-
      ment, not seriate ..................................   Sphaerophragmium (104)
5.  Cells of teliospore seriately arranged, spore head
      elongate ...........................................   Cumminsina (105)
```

1. UREDINOPSIS Magnus, Atti Congr. Internat. Bot. Genova. p.167. 1893.

Spermogonia subcuticular, type 1, 2 or 3. Aecia subepidermal in origin, erumpent, with white peridium, peridermioid; spores catenulate, verrucose. Uredinia subepidermal in origin, with peridium, opening irregularly, ostiolar cells not clearly differentiated; spores borne singly, often extruded as white tendrils, typically more or less lanceolate and apically mucronate, with a few lines of cog-like warts or smooth, germ pores near the ends; amphispores occur in some species. Telia as such not organized, the spores occur singly or in loose, not adherent groups in the mesophyll, sessile, 1-celled or usually 2- to several-celled by vertical septa, germ pore 1 in outer wall, obscure; germination occurs in overwintered fronds, basidium external.

TYPE: Uredinopsis filicina Magn.

Uredinopsis usually is considered to be the most primitive existing genus of the Uredinales, in part because the aecial states are on leaves of the genus Abies (firs) and the simple telia on ferns, and in part because no pigment is developed in the cytoplasm or the cell walls. Presumably all species are heteroecious and macrocyclic.

REFERENCES: Faull, J.H. 1938. Taxonomy and geographical distribution of the genus Uredinopsis. Contrib. Arnold Arbor. Harvard Univ. No. XI. 120 pp. Hiratsuka, N. 1958. Revision of taxonomy of the Pucciniastreae. 167 pp. Kasai Publ. and Print. Co., Tokyo. Ziller, W.G. 1974. The tree rusts of western Canada. 272 pp. Can For. Serv. Publ. No. 1329 (excellent photographic illustrations).

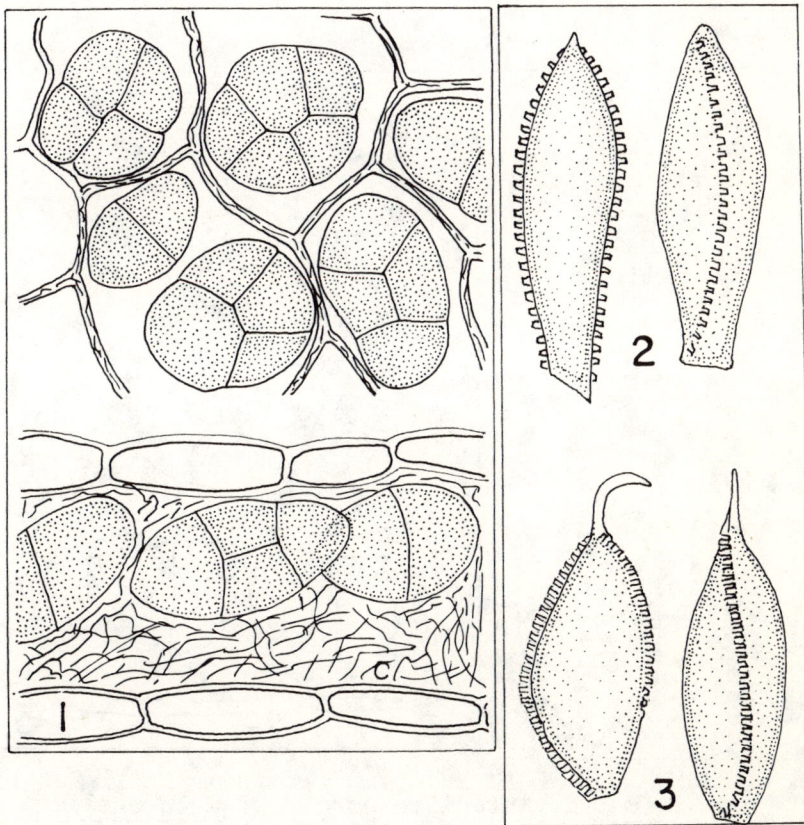

1. U. longimucronata Faull; teliospores. 2. U. pteridis Diet. & Holw.; urediniospores. 3. U. osmundae Magn.; urediniospores.

2. MILESINA Magnus, Ber. Dtsch. Bot. Ges. 27:325. 1909.

     Spermogonia subepidermal, type 1. Aecia subepidermal in origin, erumpent,
with white peridium, peridermioid; spores catenulate, verrucose. Uredinia subepidermal in origin, erumpent, with peridium, opening by an irregular pore without
clearly differentiated ostiolar cells; spores borne singly, mostly echinulate but
sometimes verrucose or smooth, pores bizonate, obscure. Telia scarcely differentiated, consisting of spores in the epidermal cells; spores few-to many-celled
by vertical septa, sessile, germ pores 1 in outer wall of each cell, obscure, the
spores of most species develop in overwintered fronds and then germinate without
dormancy, basidium external.

     LECTOTYPE: Milesina kriegeriana (Magn.) Magn. (Melampsorella kriegeriana
Magn.).

     As in Uredinopsis all spore states lack pigment and hence the sori are white.
Presumably, all species are heteroecious with aecia on the leaves of Abies (firs)
and the uredinia and telia on ferns. The teliospores are not resting spores but
develop during the winter and germinate in the spring. In addition to species of
proven lifecycle there are several putative species of Milesina that are known
only in the uredinial state. Some species produce amphispores as well as
ordinary urediniospores. Milesia has been used often but the name was based on a
uredinial (anamorphic) state.

     REFERENCES: Faull, J.H. 1932. Taxonomy and geographical distribution of the
genus Milesia. Contrib. Arnold Arbor. Harvard Univ. No. 2. 138 pp. Henderson,
D.M. and Prentice, H.T. 1977. The morphology of fungal spores: Milesina blechni
Notes R. Bot. Gard. Edinb. 35:415-417 (SEM and TEM studies). Hiratsuka, N. 1958.
Revision of taxonomy of the Pucciniastreae. 167 pp. Kasai Publ. and Print. Co.,
Tokyo.

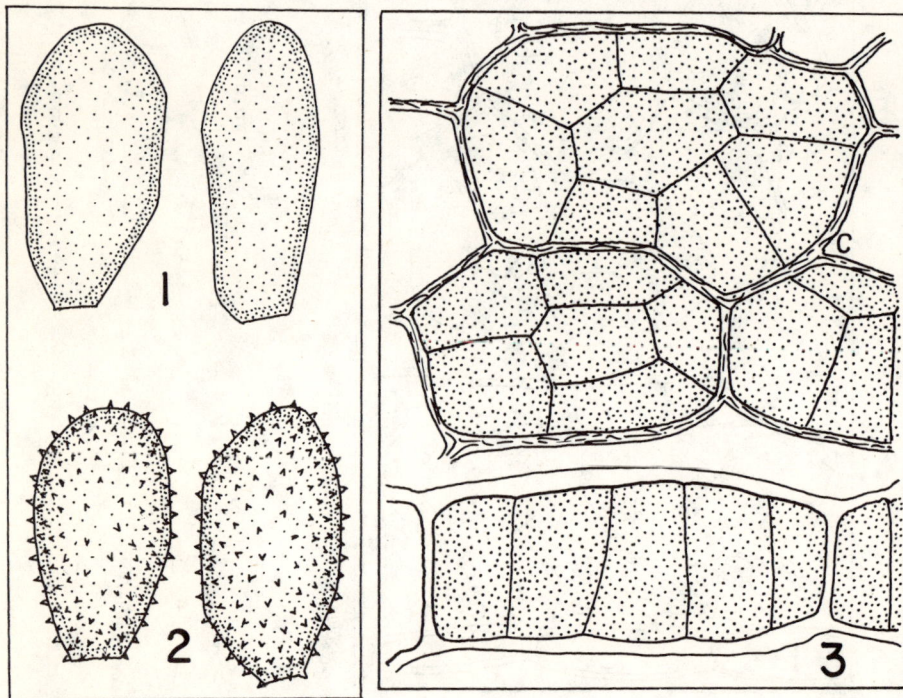

1. M. vogesiaca Syd.; urediniospores. 2. M. dieteliana
(Syd.) Magn.; urediniospores. 3. M. pycnograndis (Arth.)
Hirat. f.; telia.

3. HYALOPSORA Magnus, Ber. Dtsch. Bot. Ges. 19:582. 1901 (issued 1902).

Spermogonia subepidermal, type 2. Aecia subepidermal in origin, erumpent, with peridium, peridermioid; spores catenulate, verrucose. Uredinia subepidermal in origin, with peridium but this sometimes rudimentary and without specialized ostiolar cells, thin-walled paraphyses sometimes also present; spores borne singly, wall essentially colorless but pigment present in the cytoplasm, wall verrucose or echinulate, pores mostly scattered, obscure. Telia scarcely organized; spores produced in the epidermal cells, 2- to many-celled by vertical septa, sessile, wall thin, colorless, pore 1 in outer wall of each cell, obscure; germination occurs without dormancy, basidium external.

LECTOTYPE: Hyalopsora aspidiotus (Magn.) Magn. (Melampsorella aspidiotus Magn.).

Hyalopsora is generally similar to Milesina except that there is pigment in the cytoplasm so that the sori are yellow when fresh. But the color fades rapidly. The teliospores are not resting spores but mature in the spring on over-wintered fronds and then germinate. Amphispores are produced in addition to ordinary urediniospores and may predominate. Insofar as the lifecycles are known the species are heteroecious with aecia on Abies (firs) and uredinia and telia on ferns. Some species known only in the uredinial state probably belong here.

REFERENCES: Hiratsuka, N. 1958. Revision of taxonomy of the Pucciniastreae. 167 pp. Kasai Publ. and Print. Co., Tokyo. Ziller, W.G. 1974. The tree rusts of western Canada. 272 pp. Can. For. Serv. Publ. no. 1329 (excellent for photographic illustrations).

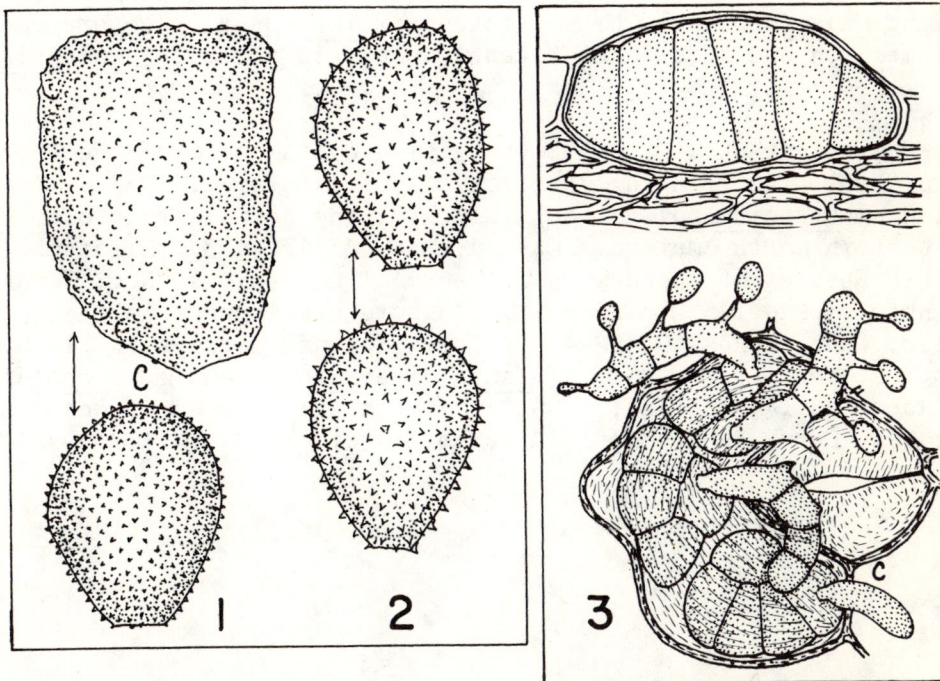

1. Amphispore (above) and one urediniospore of H. polypodii (Diet.) Magn. 2. Urediniospores of H. cheilanthis Arth.. 3. Telia of H. aspidiotus (Magn.) Magn.

4. PUCCINIASTRUM Otth, Mitt. Naturf. Ges. Bern 1861:71. 1861.

Spermogonia subcuticular, type 3. Aecia subepidermal in origin, erumpent, with peridium, peridermioid; spores catenulate and verrucose or uredinoid with spores borne singly. Uredinia subepidermal in origin, with peridium, opening by a pore delimited by differentiated ostiolar cells; spores borne singly, wall colorless, echinulate, pores scattered, obscure. Telia 1 spore deep, consisting of subepidermal crusts of laterally adherent spores (Pucciniastrum sensu stricto) or of laterally adherent spores within epidermal cells of the host (Thekopsora and Calyptospora, when segregated); spores sessile, 2- or more-celled by vertical septa, germ pore 1 in the outer wall of each cell, wall pigmented; germination occurs after dormancy, basidium external.

TYPE: Pucciniastrum epilobii Otth.

All species whose life cycle is known are heteroecious, with aecia on leaves of Abies, Picea, or Tsuga, and the uredinia and telia on dicotyledonous plants, including the Rosaceae and Ericaceae. There are species of Uredo that will prove to belong in Pucciniastrum; one of these is on Orchidaceae. P. goeppertianum (Kuehn) Kleb., lacks uredinia and often is treated as a separate genus, Calyptospora. Uraecium holwayi (Arth.) Arth., on needles of Tsuga, produces uredinia and telia on Vaccinium membranaceum in western North America (Hiratsuka, Y., 1965). P. vaccinii (Wint.) Joerst. thus has a western form different from the more widely distributed species with peridermioid aecia. Species with uraecia occur also in Japan (Hiratsuka, Y. and Sato, 1976, Sato and Katsuya, 1979).

Thekopsora and Calyptospora have commonly been treated as separate genera because of the position of the teliospores and, to some degree, because of the mode of formation (see Pady). There is logic to recognizing Thekopsora but none for considering Calyptospora to be separate from Thekopsora. We are continuing the practice of American and some other recent manuals in accepting Pucciniastrum sensu lato.

REFERENCES: Hiratsuka, N. 1958. Revision of taxonomy of the Pucciniastreae, 167 pp. Kasai Publ. and Printing Co., Tokyo. Hiratsuka, Y. 1965. The identification of Uraecium holwayi on Hemlock as the aecial state of Pucciniastrum vaccinii in Western North America. Can. J. Bot. 43:475-478. Hiratsuka, Y. and Sato, S. 1976. Species of Thekopsora on Tsuga. Trans. Mycol. Soc. Japan 17:543-548 (in Japanese). Pady, S.M. 1933. Teliospore development in the Pucciniastreae. Can. J. Res. 9:458-485. Sato, S., and Katsuya, K. 1979. Heteroecism of two rust fungi on needles of Tsuga diversifolia and T. sieboldii. Trans. Mycol. Soc. Japan 20:1-4. Ziller, W.G. 1974. The tree rusts of western Canada. 272 pp. Can. For. Serv. Publ. No. 1329 (excellent for photographic illustration).

Illustration next page.

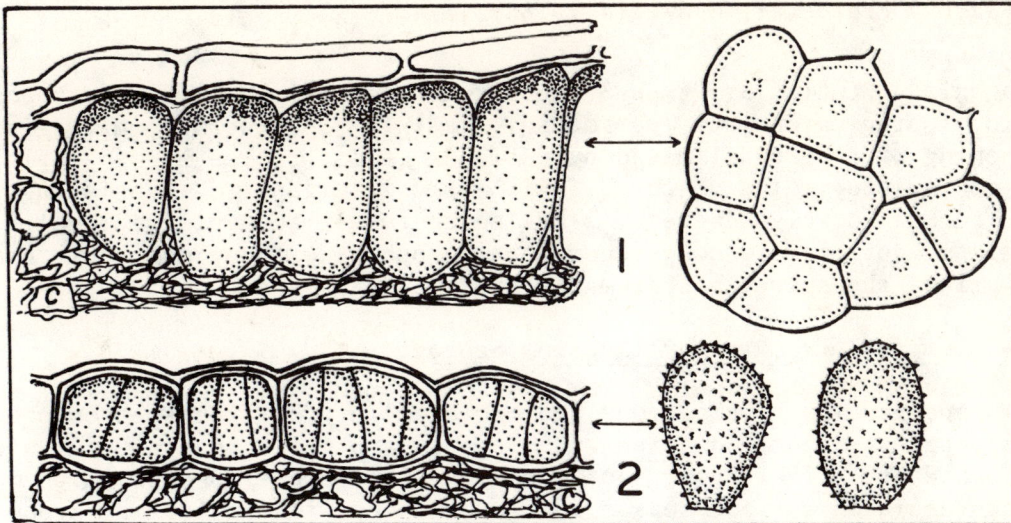

1. P. epilobii Otth; teliospores in section and surface view.
2. P. vaccinii (Wint.) Joerst.; intraepidermal teliospores and urediniospores.

5. MELAMPSORELLA Schroeter, Hedwigia 13:85. 1874.

Spermogonia subcuticular, type 3. Aecia subepidermal in origin, erumpent, peridermioid; spores catenulate, verrucose. Uredinia subepidermal in origin, with peridium opening by a pore delimited by ostiolar cells; spores borne singly, wall colorless, echinulate, pores obscure. Telia scarcely organized, consisting of closely packed but loosely adherent, 1-celled spores in the epidermal cells, wall thin, colorless, germ pore obscure, germination occurs without dormancy through the outer wall of the spore, basidium external.

TYPE: Melampsorella caryophyllacearum Schroet.

The two species are heteroecious and macrocyclic. The type species produces conspicuous witches' brooms on Abies and uredinia and telia on systemic infections of Cerastium and Stellaria. Reports of aecia on Picea are erroneous. M. symphyti Bub. produces aecia on current season leaves of Abies without forming brooms, and uredinia and telia on Symphytum (Boraginaceae). Neither species has resting teliospores.

REFERENCES: Hiratsuka, N. 1958. Revision of taxonomy of the Pucciniastreae. 167. pp. Kasai Publ. and Print. Co., Tokyo. Pady, S.M. 1946. The development and germination of the intraepidermal teliospores of Melampsorella cerastii. Mycologia 38:477-499.

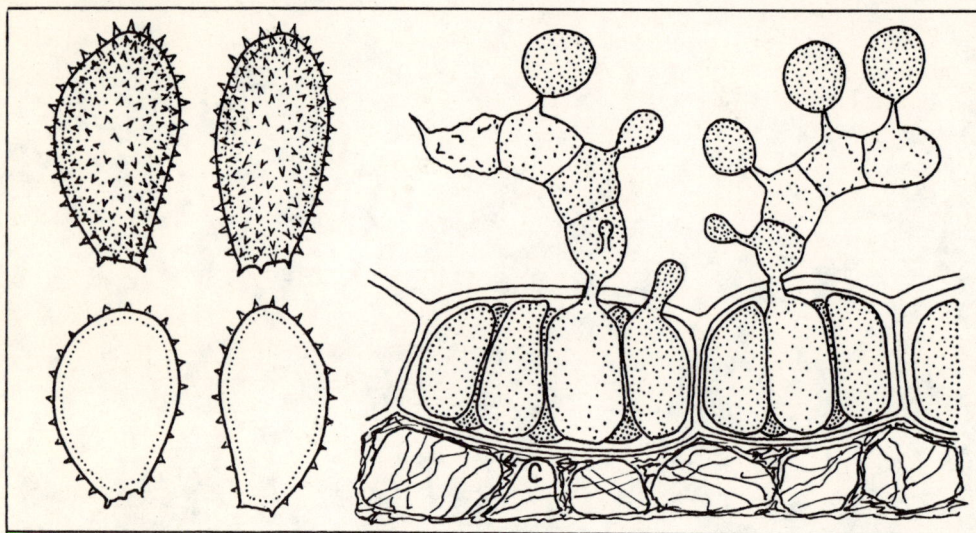

M. caryphyllacearum Schroet.; urediniospores and germinating teliospores.

6. MELAMPSORIDIUM Klebahn, Z. Pflanzenkr. 9:21. 1899.

Spermogonia subcuticular, type 3. Aecia subepidermal in origin, erumpent, with peridium, peridermioid; spores catenulate, verrucose. Uredinia subepidermal with peridium opening by a discrete pore delimited by differentiated ostiolar cells; spores borne singly, echinulate, wall colorless, pores bizonate, obscure. Telia subepidermal, not erumpent, consisting of crusts 1 spore deep of closely packed spores; spores sessile, 1-celled, with thin colorless walls, germ pore apical if differentiated, germination occurs after wintering on fallen leaves, basidium external.

TYPE: Melampsoridium betulinum (Fries) Kleb. (Sclerotium betulinum Fries).

Melampsoridium is a genus of four heteroecious and presumably heteroecious species. The aecia of M. betulinum occur on needles of Larix and the telia on species of Alnus and Betula. The ostiolar cells of the uredinial peridium, because of the long spine-like apex of most species, makes such species easy to recognize when uredinia are present, and they are the obvious spore state. M. inerme, on Magnolia , is described as not having the spinescent ostiolar cells. Telia are not often collected because they develop late in autumn. F. & H. Roll-Hansen suggest, from inoculation and morphology, that M. alni (Thuem.) Diet. is synonymous with M. betulinum, but Kaneko and Hiratsuka report that there are differences in urediniospore germ pores.

REFERENCES: Hiratsuka, N. 1958. Revision of taxonomy of the Pucciniastreae. 167. pp. Kasai Publ. & Print. Co., Tokyo. Roll-Hansen, F.& H. 1981. Melampsoridium on Alnus in Europe. M. alni conspecific with M. betulinum. Eur. J. For. Path. 11: 77-87. Kaneko, S. and Hiratsuka, N. 1981. Classification of the Melampsoridium species based on the position of urediniospore germ pores. Trans. Mycol. Soc. Japan. (in Japanese) 22:463-473. Singh, S. and Pandy, P.C. 1972. Melampsoridium inerme on Magnolia. Trans. Brit. Mycol. Soc. 58:342-344.

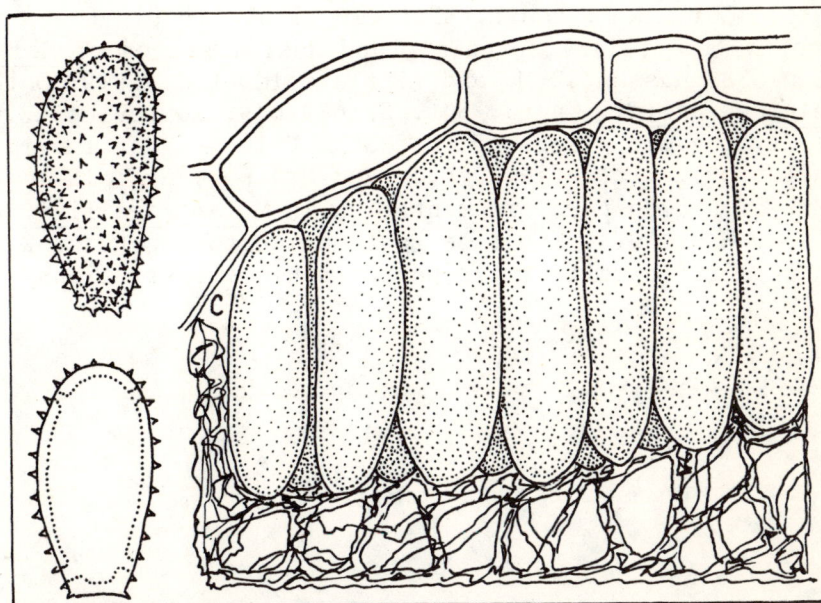

M. betulinum (Fries) Kleb.; urediniospores and a telium.

7.   MELAMPSORA Castagne, Obs. Pl. Acotyl. Fam. Ured. 2:18. 1843.

Spermogonia subcuticular, type 3, or subepidermal, type 2. Aecia subepidermal in origin, erumpent, generally considered to be caeomatoid but some species have peridial cells adherent to the host epidermis; spores catenulate, verrucose with rod-like columns or blocks. Uredinia subepidermal in origin, erumpent, bright yellow or orange when fresh, fading to nearly colorless, with abundant capitate paraphyses and in some species a partial peridium also; spores borne singly on pedicels, wall colorless, echinulate, pores scattered or bizonate, obscure. Telia subepidermal or rarely subcuticular, remaining covered, consisting of laterally adherent crusts 1 spore deep; spores 1-celled, sessile (some species have spore-like, presumably sterile, cells below the spores) wall brown or brownish, pore 1, basidium external.

TYPE: _Melampsora euphorbiae_ (Schub.) Cast. (_Xyloma euphorbiae_ Schub.).

Numerous species have been described but not all are distinct morphologically. Both heteroecious and autoecious species occur. Most heteroecious species produce aecia on the needles of conifers and telia on _Populus_ and _Salix_, but some produce the aecia on _Allium_, _Saxifraga_, or _Ribes_. Some species are not obligatorily heteroecious and persist far from the putative aecial hosts. The autoecious species occur on various dicotyledonous families, including Euphorbiaceae and Linaceae. _M. farlowii_ (Arth.) J.J. Davis is microcyclic on _Tsuga canadensis_ in North America. _M. hypericorum_ Wint., on _Hyperium_, supposedly is demicyclic but lacks spermogonia. But the aecia (?) have peridial cells adherent to the up-turned epidermis and verrucose, catenulate spores and intercalary cells. The species has, at times, been segregated as _Mesopsora hypericorum_ (Wint.) Diet.
Melampsora lini (Ehrenb.) Lév. is agriculturally the most important pathogen but where poplar trees are economically important, defoliation by several species of _Melampsora_ may be severe enough to warrant control measure.

REFERENCES: No monograph, other than vol. 3 of Sydow's Monographia Urediniarum, exists. Gold, R.E. and Littlefield, L.J. 1979. Light and scanning electron microscopy of the telial, pycnial, and aecial stages of _Melampsora lini_. Can. J. Bot. 57:629-638. Hassan, Z.M. and Littlefield, L.J. 1979. Ontogeny of the uredium of _Melampsora lini_. Can J. Bot. 57:639-649. Kuprevich, V.F. and Tranzschel, V.G. 1957. Cryptogamic plants of the USSR. Vol. 4. Rust fungi. No. 1. 518 pp. (English translation). Littlefield, L.J. and Bracker, C.E. 1971. Ultrastructure and development of urediospore ornamentation in _Melampsora lini_. Can. J. Bot. 49:2067-2073. Ziller, W.G. 1974. The tree rusts of western Canada. 272 pp. Can. For. Serv. Publ. No. 1329 (excellent for photographic illustrations).

Illustrations next page.

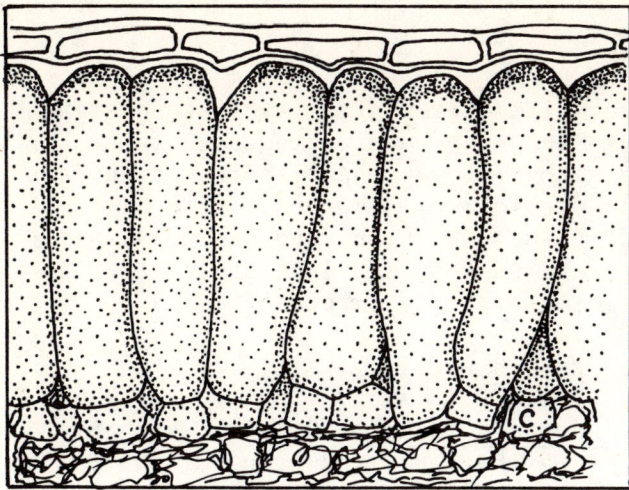

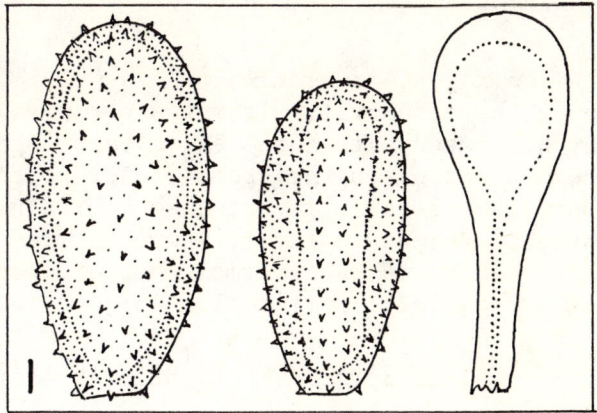

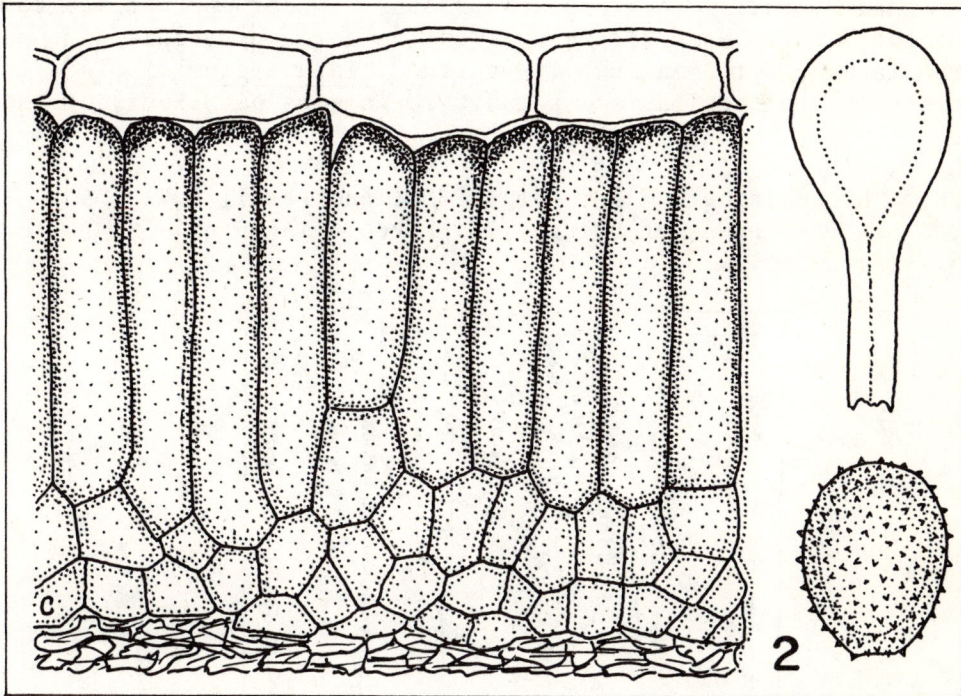

1. M. *occidentalis* H.S. Jack.; telium and urediniospores.
2. M. *monticola* Mains; telium, uredinial paraphysis, and urediniospores.

8.   HIRATSUKAMYCES Thirumalachar, Kern & Patil, Sydowia 27:80. 1975.

Spermogonia subepidermal, type 12. Aecia subepidermal, 3 or 4 cells
deep in the mesophyll, without a peridium but with "dermatate" layers consisting
of hyphae fused with host cells, becoming exposed by extensive rupture of the
epidermis and extruding spores interspersed with elater hyphae; spores catenulate
but separating easily, verrucose. Uredinia subepidermal, erumpent, with a
peripheral dermatate layer; spores produced in succession on long pedicels, verrucose. Telia intraepidermal, not erumpent; spores sessile, 1-celled, 3 to 5 in
an epidermal cell; germination unknown, basidium probably external.

TYPE: Hiratsukamyces salacicola Thir., Kern & Patil.

Two species have been recognized plus Caeoma callianthum H. Syd. which probably belongs to the genus. The aecial states may be localized or cover the entire
surface of leaves causing first an elevation of the host tissue in somewhat cere-
broid patterns then rupturing to expose the spores. Under appropriate conditions
the spores together with the elaters hang from the sorus. Because of these unique
features Thirumalachar, Kern, and Patil described Elateraecium as a new aecial
type. The species are autoecious and parasitize species of Salacia of the
Celastraceae.

REFERENCE: Thirumalachar, M.J., Kern, F.D., and Patil, B.V. 1966.
Elateraecium, a new form genus of Uredinales. Mycologia 58:391-396 (with photographs of dehiscing aecia).

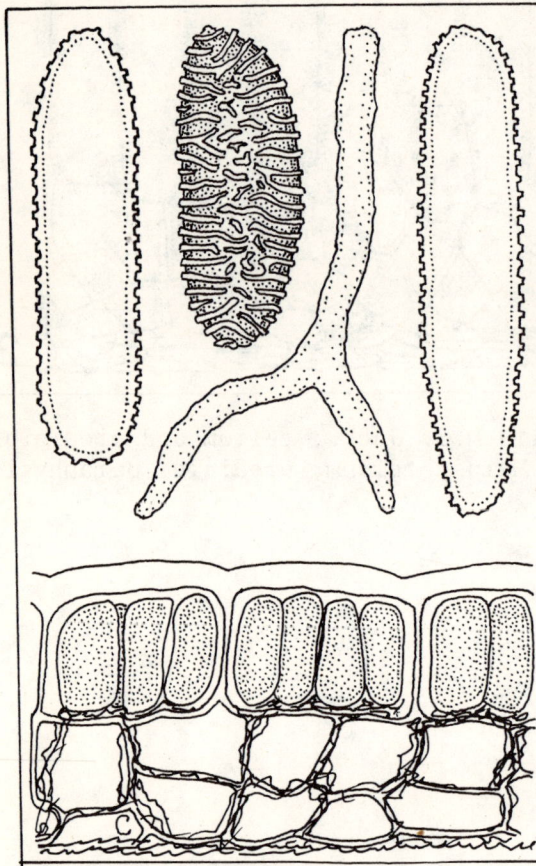

H. salicicola Thir., Kern & Patil;
three aeciospores, an elater hypha
and intraepidermal teliospores.

9.   PHAKOPSORA Dietel, Ber. Dtsch. Bot. Ges. 13:333. 1895.

Spermogonia subcuticular, type 7. Aecia subepidermal in origin, erumpent, uredinoid; spores borne singly and similar to the urediniospores. Uredinia subepidermal in origin, erumpent, with peripheral, incurve, usually dorsally thick-walled paraphyses surmounting peridial tissue (Phakopsora sensu stricto), or without (Bubakia when segregated); spores borne singly, wall echinulate, brownish or nearly colorless , pores scattered or equatorial.Telia subepidermal, not erumpent, consisting of crusts of laterally adherent spores 2 or more cells deep, sessile but not catenulate, irregularly arranged, 1-celled, the wall usually brown or brownish, with 1 apical germ pore, presumably germinating after dormancy in most or all species, basidium external.

TYPE: Phakopsora punctiformis (Diet. & Barcl.) Diet. (Melampsora punctiformis Diet. & Barcl.).

Phakopsora differs from Physopella because the teliospores are irregularly arranged whereas those of Physopella are in chains, but this difference may not be distinct. The uredinia of both genera commonly have paraphyses but there are species in each genus that have none. The species that often are segregated in Bubakia are autoecious. Spermogonia and aecia are not known for Phakopsora sensu stricto and, because numerous species are known, it is possible that some or all may be heteroecious,
There are 50 or more species on monocots and dicots. Most are distributed in warmer region. Numerous Uredo species also may belong here. Phakopsora pachyrhizae H. Syd. & P. Syd. (Physopella pachyrhizae (Syd.) Azbu.) on soybean and P. gossypii (Lagh.) Hirat. f. on cotton are of economic importance.

REFERENCES: Azbukina, Z.M. 1970. O sistematicheskom polozhenii i geneticheskikh rodov triby Phakopsoreae Arth. emend. Azb. Nov. Sist. niz. Rast. 7:208-232. Hiratsuka, N. 1935, 1936. Phakopsora of Japan. I. Bot. Mag. Tokyo 49:781-788; II. ibid. 49:853-860; III. ibid. 50:2-8. Thirumalachar, M.J. and Kern, F.D. 1949. Notes on some species of Phakopsora and Angiopsora. Mycologia 41:283-290.

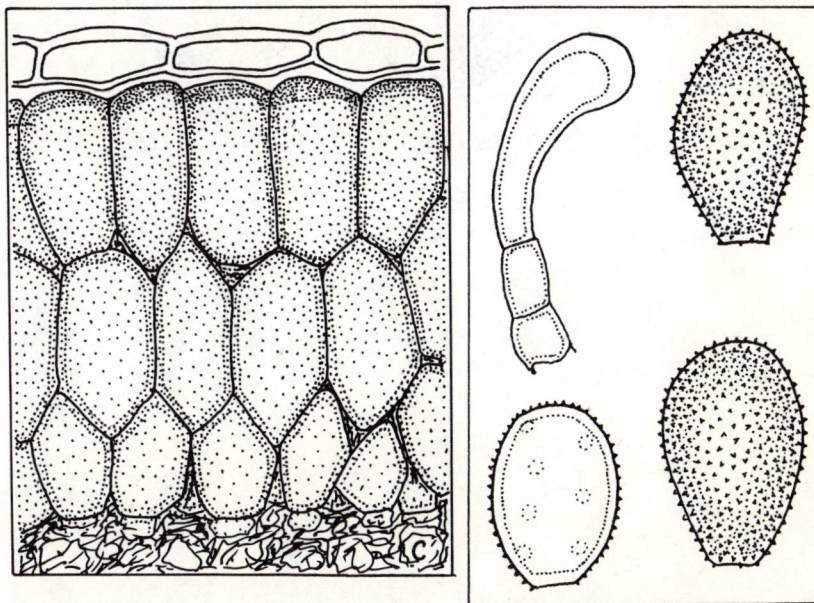

P. pachyrhizae Syd.; telia from Desmodium and one paraphysis and spores from uredinia on soybean.

35

9E.   MONOSPORIDIUM Barclay, J. Asiatic Soc. Bengal 56:367. 1887.

    Spermogonia subcuticular. Aecia and uredinia not produced. Telia subepidermal in origin, erumpent, with peridium, resembling Aecidium; spores 1-celled, catenulate, basidium external.

    LECTOTYPE: Monosporidium andrachnis Barcl.

    Monosporidium is an endo type "genus" differing from Endophyllum in having type 7 (or 5?) spermogonia. It has the appearance of Aecidium but the spores produce basidia and basidiospores rather than vegetative mycelium. Monosporidium is, thus, a microcyclic life cycle variant of a macrocyclic species, whether the parental form is known or not. In theory such species could derive from any genus that has type 7 (or 5?) spermogonia and aecidioid aecia. Kulkarniella is synonymous.

    REFERENCES: Gokhale, V.P. and Patil, M.K. 1951 (issued 1952).Kulkarniella, a new genus of rusts. Indian Phytopathol. 4:171-173. Thirumalachar, M.J. and Kern, F.D. 1955. The rust genera Allotelium, Atelocauda, Coniostelium and Monosporidium. Bull. Torrey Bot. Club 82:102-107.

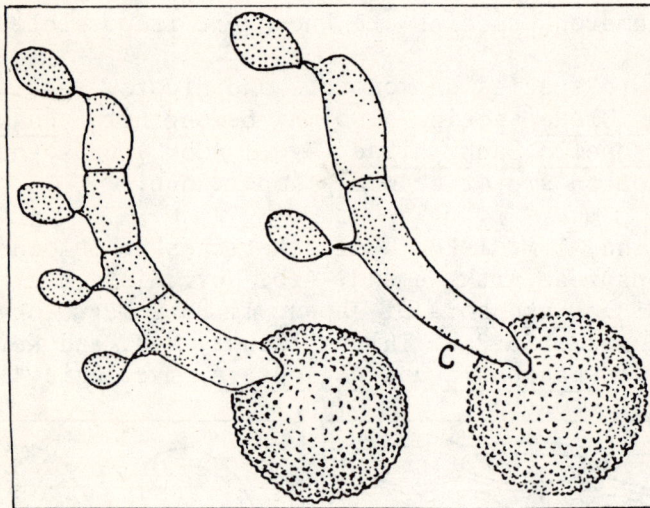

Monosporidium; teliospores; schematic.

10. PHYSOPELLA Arthur, Rés. Sci. Congr. Internat. Bot. Vienna. p. 338. 1906.

Spermogonia subcuticular, type 7. Aecia subepidermal in origin, erumpent, with peridium, aecidioid; spores catenulate, verrucose. Uredinia subepidermal in origin, becoming erumpent, mostly with peripheral, incurved, dorsally thick-walled, basally septate (or surmounting peridial tissue) paraphyses; spores borne singly, echinulate, pores scattered or equatorial. Telia subepidermal, not erumpent, consisting of laterally adherent spores forming crusts 2 or more spores deep; spores sessile, catenulate, 1-celled, germ pore 1, apical, obscure, wall pigmented, germination occurs after dormancy (in all?), basidium external.

TYPE: Physopella ampelopsidis (Diet. & Syd.) Cumm. & Rama. (Phakopsora ampelopsidis Diet. & Syd.)

There are some 20 or more species known on monocots and dicots; most inhabit warm regions. Several species of Uredo may prove to belong to Physopella. Type 7 spermogonia and aecidioid aecia have been ascribed to P. hansfordii (Cumm.) Cumm. & Rama. and have been illustrated in P. ampelopsidis by Hiratsuka and Sato.

REFERENCES: Azbukina, Z.M. 1970. O sistimaticheskom polozhenii i genetich-eskikh svyazakh rodov triby Phakopsoreae Arth. emend. Azb. Nov. Sist. Niz. Rast. 7:208-232. Cummins, G.B. and Ramachar, P. 1958. The genus Physopella replaces Angiopsora. Mycologia 50:741-744. Hiratsuka, Y. and Sato, S. 1982. Morphology and taxonomy of rust fungi, p.1-36. In Scott, K.T. and Chakravarty, A.K., eds. The rust fungi. Academic Press. 288 p. Thirumalachar, M.J. and Kern, F.D. 1949. Notes on some species of Phakopsora and Angiopsora. Mycologia 41:283-290.

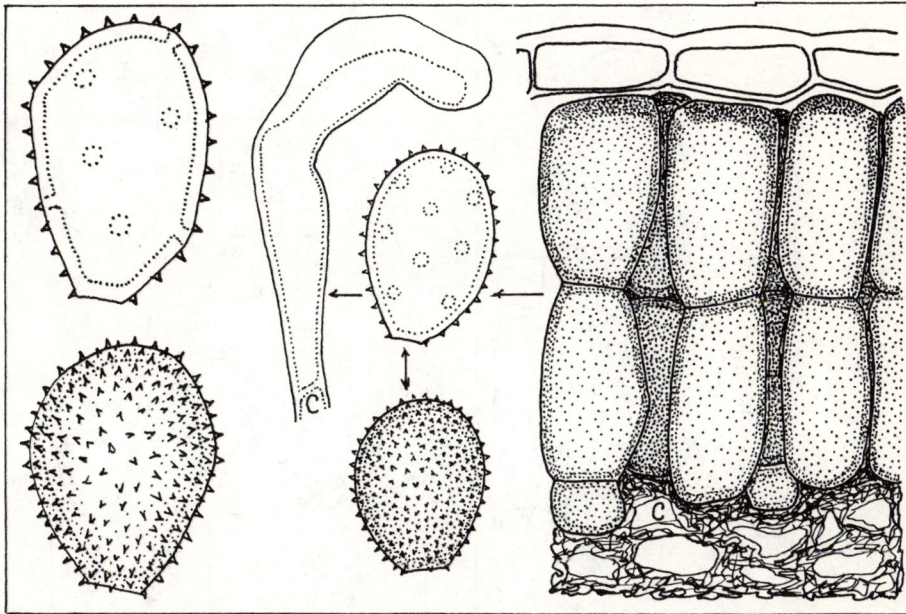

P. mexicana Cumm.; urediniospores (left); P. compressa (Mains) Cumm. & Ramachar, paraphysis, two urediniospores, and teliospores (right).

11.   PUCCINIOSTELE Tranzschel & Komarov, Arb. Nat. Ges. St. Peters. 30:138. 1899.

Spermogonia subcuticular, type 7. Aecia subepidermal in origin, erumpent, without peridium, caeomatoid; spores catenulate, verrucose. Uredinia subepidermal in origin, erumpent, with peridium and catenulate spores or with peripheral paraphyses and spores borne singly. Telia subepidermal in origin, only slowly exposed; spores catenulate, of 2 kinds: (1) the primary spores develop in aecia or closely associated and consist of catenulate, 4-celled, tetrad-like disks, and (2) secondary spores which develop in separate sori, are 1-celled and catenulate, germpores not seen, germination unknown but basidium doubtless external.

TYPE: Pucciniostele clarkiana (Barcl.) Tranz. & Kom. (Xenodochus clarkiana Barcl.).

Pucciniostele is a poorly understood genus and anomalous in that it apparently produces two kinds of teliospores, but neither kind has been germinated. There are three species that parasitize Astilbe of the Saxifragaceae and P. hashiokai (Hirat. f.) Cumm. on Ampelopsis (Vitaceae). The latter has paraphysate uredinia with pedicellate spores and teliospores that are unlike those of other species. The teliospores resemble those of Phragmidium but without pedicels. It is questionable that the fungus belongs to any described genus. Hiratsuka described it as a species of Cerotelium but Sawada named it in Pucciniostele.

REFERENCE: Cummins, G.B. and Thirumalachar, M.J. 1953. Pucciniostele a genus of the rust fungi. Mycologia 45:572-578.

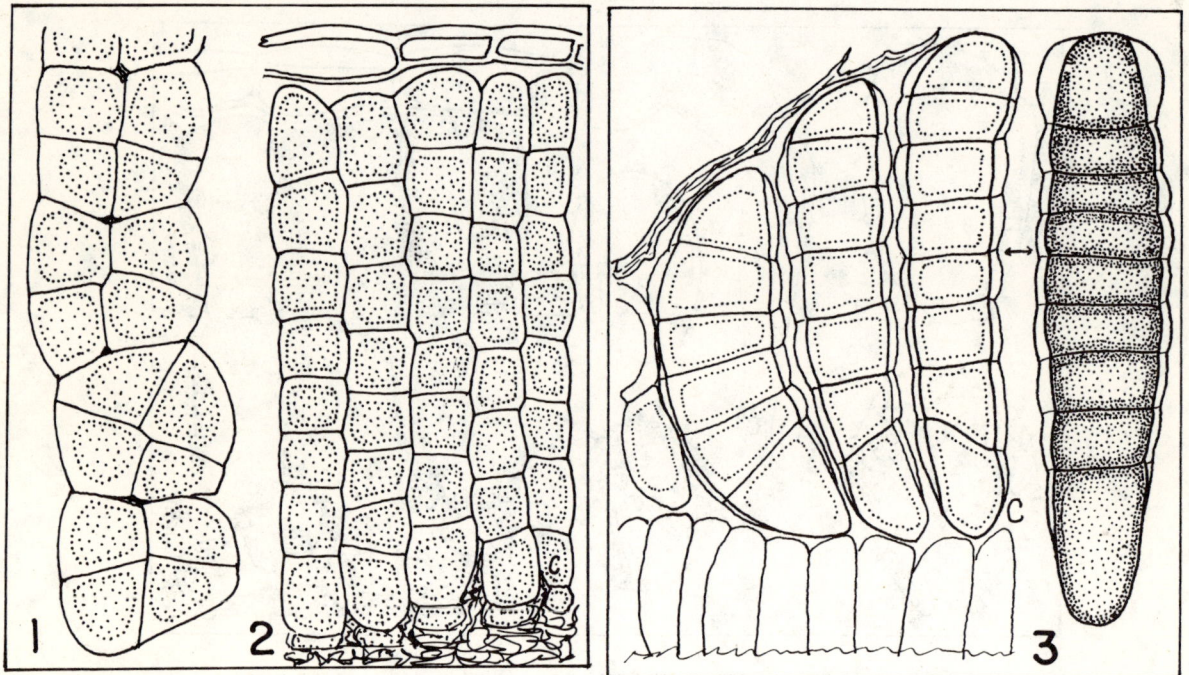

1. P. clarkiana (Barcl.) Tranz. & Kom.; tetrad-like teliospores. 2. P. mandshurica Diet.; catenulate teliospores. 3. P. hashiokai (Hirat. f.) Cumm.; This species may not belong in the genus.

12. COLEOSPORIUM Léveillé, Ann. Sci. Nat. III ser. 8:373. 1847.

Spermogonia subepidermal, type 2. Aecia subepidermal in origin, erumpent, with conspicuous peridium, peridermioid; spores catenulate, verrucose with rod-like columns or knobs, some with annulate sides, or broadly conical. Uredinia subepidermal, erumpent, bright orange when fresh, fading whitish; spores catenulate, verrucose or echinulate or reticulate in some species, generally as the aeciospores of the species, pores scattered, obscure. Basidiosori subepidermal in origen, erumpent as low or rarely columnar cushions, hard when dry, gelatinous wet; basidia sessile, 1-celled, in 1-layered crusts, or pseudocatenulate by intrusion of young spores among older spores, or catenulate, wall thick and gelatinizing above, germination occurs without dormancy by division of the protoplast into a 4-celled basidium, each cell producing a sterigma and one basidiospore.

LECTOYPE: Coleosporium campanulae (Strauss) Tul.

Coleosporium has many described species, some of which are doubtfully distinct morphologically, but SEM studies (Hiratsuka and Kaneko, 1975) of the surface sculpture of aeciospores and urediniospores show greater variability than previously recognized. Further, Kaneko (1970) has recognized two types of basidial development. In type 1 the entire cell becomes a basidium without change in length; in type 2 when the cell matures a septum forms that divides the upper basidial part from a basal stalk-like part. Basidiospore shape varies also (Hiratsuka and Kaneko, 1977; Cummins, 1978), being more or less napiform-ellipsoid in some species but oblong-ellipsoid or nearly allantoid in some. Any monograph of the genus doubtless will use these characteristics (see Kaneko, 1981).

Most species are heteroecious, with the spermogonia and aecia on needles of Pinus and basidia on monocotyledonous and dicotyledonous plants. There are more species on Compositae than on any other family. Most species are macrocyclic but there are life cycle variants. There are three demicyclic, autoecious species on Compositae, C. incompletum Cumm., C. reichei Diet., and C. viguierae Diet. & Holw. and four microcyclic species on Pinus, C. crowellii Cumm., C. pinicola (Arth.) Arth., C. himalayense Durr., and C. pini-pumilae Azbu. These species follow Tranzschel's Law in that they produce basidia on the aecial host of macrocyclic species. Arthur treated C. pinicola as a derivative of C. inconspicuum Arth. Those who believe that microcyclic species should be segregated in a distict genus because of catenulate basidia overlook the occurrence of catenulate basidia in some macrocyclic species. The relationship of the microcyclic species is obvious.

REFERENCES: Cummins, G.B. 1978. Rust Fungi on legumes and composites in North America. Univ. Arizona Press. 424. pp. Henderson, D.M. and Prentice, H.T. 1974. Spore morphogenesis of Coleosporium tussilaginis. Trans. Brit. Mycol. Soc. 63: 431-435. Hiratsuka, N. and Kaneko, S. 1975. Surface structure of Coleosporium spores. Rept. Tottori Mycol. Inst. 12:1-13. Hiratsuka, N. and Kaneko, S. 1977. Significance of some morphological characters of telial state in taxonomy of Coleosporium. Abstr. 2nd. Internat. Mycol. Congr. Vol. A-L;289. Holm, L., Dunbar, A., and Hofsten, A. 1970. Studies on the fine structure of aeciospores. Svenska Bot. Tidskr. 64:380-382. Kaneko, S. 1977. Coleosporium pedunculatum, a new species of pine needle rusts. Rept. Tottori Mycol. Inst. 15:13-21. Kaneko, S. 1981. The species of Coleosporium, the causes of pine needle rusts, in the Japanese Archipelago. Rept. Tottori Mycol. Inst. 19:1-159. Kuprevich, V.F. and Tranzschel, V.G. 1957. Cryptogamic plants of the USSR. Rust fungi No. 1. 518 pp. (English translation, 1970). Saho, H. 1968. Studies on the needle rusts of five-needle pines. (Japanese with English summary) Bull. Tokyo Univ. Forests. 64:59-148.

Illustrations next page.

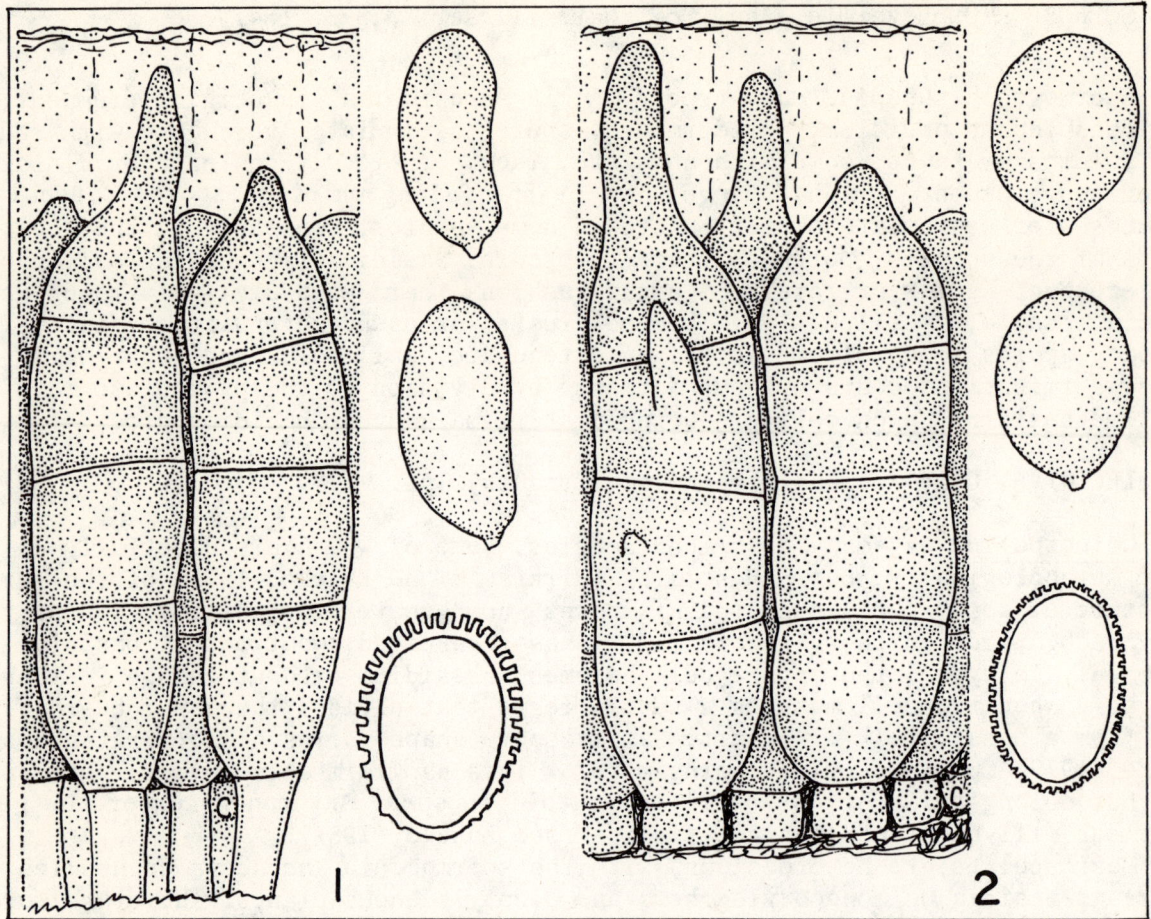

1. C. anceps Diet. & Holw.; teliospores, two basidiospores, and one uredinious pore. 2. C. sonchi (Strauss) Tul.; teliospores, two basidiospores, and one urediniospore. The differing bases and shapes of basidiospores have been used for specific distinctions by Kaneko, 1981.

13. GOPLANA Raciborski, Paras. Algen Pilze Javas 2:24. 1900.

Spermogonia and aecia unknown. Uredinia subepidermal in origin, erumpent; spores borne singly, pores obscure, superequatorial in 1 species. Basidiosori subepidermal in origin, erumpent as gelatinous (when wet) cushions; basidia in a single layer, sessile, embedded in a gelatinous matrix, becoming 4-celled by division of the protoplast, each cell produces a basidiospore; germination occurs without dormancy.

TYPE: Goplana micheliae Racib.

Some 10 species are known, all of which occur in the warmer regions of the world. The type species is on Magnoliaceae. Most are represented by few collections; the life history is known for none. The pedicellate urediniospores distinguish Goplana from Coleosporium, with which it probably has no relationship.

REFERENCE: Ono, Y. 1978. Taxonomic revisions of the tribe Oliveae and morphologically related genera (Uredinales). Ph.D. thesis Purdue University 254 pp.

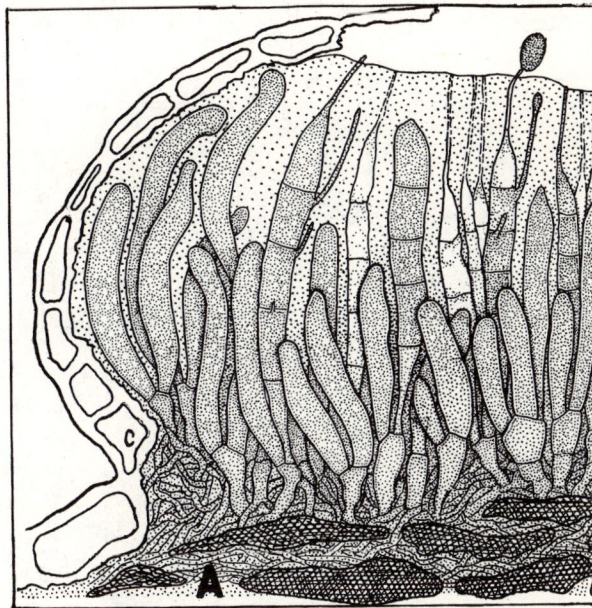

G. dioscoreae Cumm.; basidiosorus.
From Cummins, Mycologia 1935.

14. OCHROPSORA Dietel, Ber. Dtsch. Bot. Ges. 13:401. 1895.

Spermogonia subcuticular, type 7. Aecia subepidermal in origin, erumpent, with peridium, aecidioid; spores catenulate, verrucose. Uredinia subepidermal in origin, erumpent, the type species with abundant, peripheral, paraphyses joined below to peridial tissue; spores borne singly on short pedicels, echinulate, pores obscure. Basidiosori subepidermal in origin, in crusts 1 cell deep, waxy in appearance, becoming erumpent; basidia 1-celled becoming 4-celled by division of the protoplast, each cell produces 1 basidiospore, wall thin and pale; germination occurs without dormancy.

TYPE SPECIES: Ochropsora ariae (Fckl.) Syd. (O. sorbi (Oud.) Diet.)

There are 3 species. O. ariae and O. kraunhiae (Diet.) Diet. are heteroecious producing aecia on Anemone or Corydalis, respectively. Both aecial states are systemic. Because none of the spore states are "resting" spores, systemic infections or some other compensatory development would be expected.

REFERENCE: Hiratsuka, N. and Kaneko, S. 1978. Heteroecism of the Wistaria rust. Ochropsora kraunhiae (Diet.) Diet. Proc. Japan. Acad. 54(ser. B), No. 6:300-303.

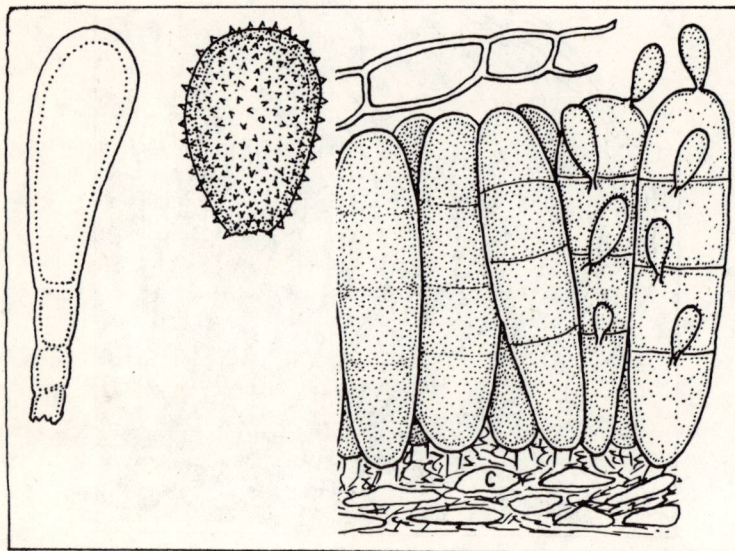

O. ariae (Fckl.) Syd.; one uredinial paraphysis, one urediniospore, and teliospores.

15. MIKRONEGERIA Dietel in Dietel & Neger, Bot. Jahrb. 27:16. 1899.

Spermogonia deep-seated, usually lobed or multiloculate (type 12). Aecia subepidermal in origin, erumpent, caeomatoid; spores catenulate, verrucose. Uredinia subepidermal in origin, erumpent; spores borne singly on pedicels, echinulate, pores obscure or not differentiated. Telia subepidermal in origin, erumpent, waxy; spores 1-celled, sessile, laterally free, without a differentiated germ pore, germination occurs without dormancy by continued growth of the apex of the spore, basidium external.

TYPE: Mikronegeria fagi Diet. & Neger.

Two species have been described. M. fagi produces telia on Nothofagus and aecia on Araucaria. M. alba R. Pete & E. Oehr. also produces telia on Nothofagus but aecia on Austrocedrus.

REFERENCES: Butin, H. 1968. Studien zur Morphologie und Biologie von Mikronegeria fagi Diet. & Neger, Phytopathol. Z. 64:242-257. Peterson, R.S. 1968. Rust fungi on Araucariaceae. Mycopathol. Mycol. Appl. 34:17-26. Peterson, R.S. & Oehrens, E. 1978. Mikronegeria alba (Uredinales). Mycologia 70:321-331.

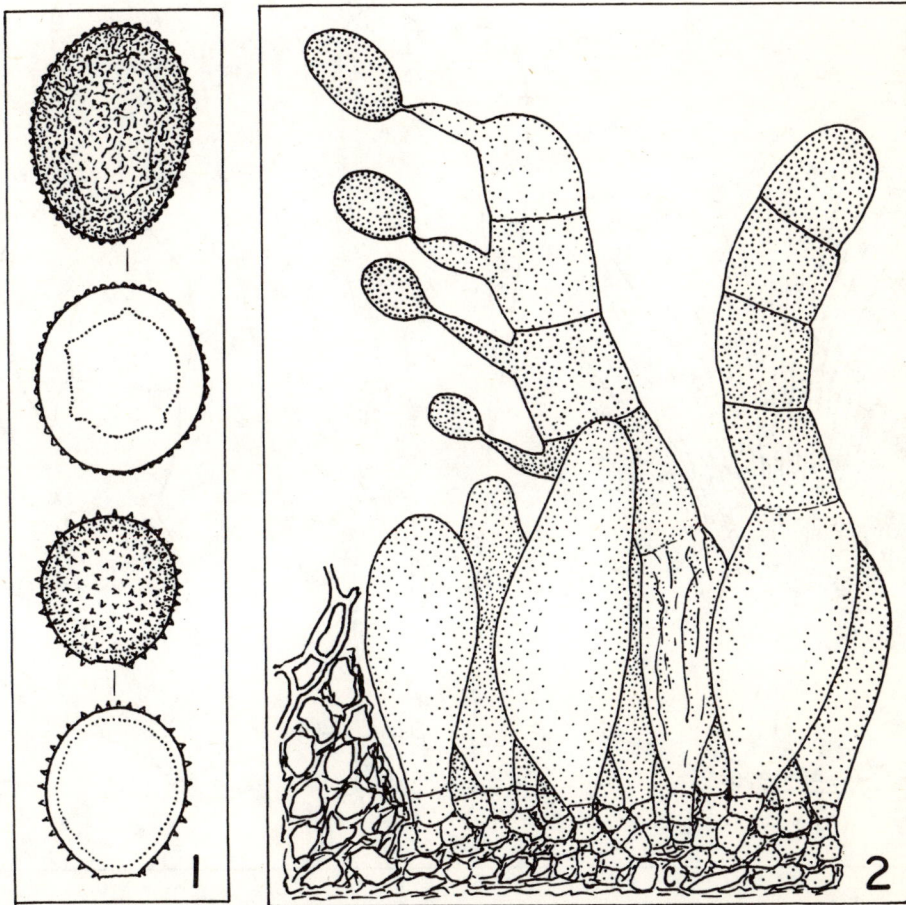

1. M. alba R. Pete. & E. Oehr.; aeciospores (above) and urediniospores. 2. M. fagi Diet. & Neger; germinating teliospores.

16. CEROPSORA Bakshi & Singh, Can. J. Bot. 38:260. 1960.

Spermogonia, aecia, and uredinia unknown. Telia subepidermal in origin, becoming erumpent as waxy cushions; spores 1-celled, borne singly on compacted, parallel, multiseptate hyphae, the spores laterally adherent when young but separating later, the wall colorless, without a differentiated germ pore, germination occurring without dormance by continued elongation of the apex of the spore; basidium external.

TYPE: Ceropsora piceae (Barcl.) Bakshi & Singh (Chrysomyxa piceae Barcl.).

The species occurs in India on the Himalayan spruce (Picea smithiana (Wall.) Boiss. and is not recorded from elsewhere. There are ill-defined peripheral cells which perhaps represent a vestigial peridium. The species doubtless is microcyclic.

Ceropsora has robust teliospores, broad basidia, and large basidiospores and, in these features, is similar to Mikronegeria, but it differs because of the extensive cellular base below the teliospores.

Bakshi and Singh published excellent photographs.

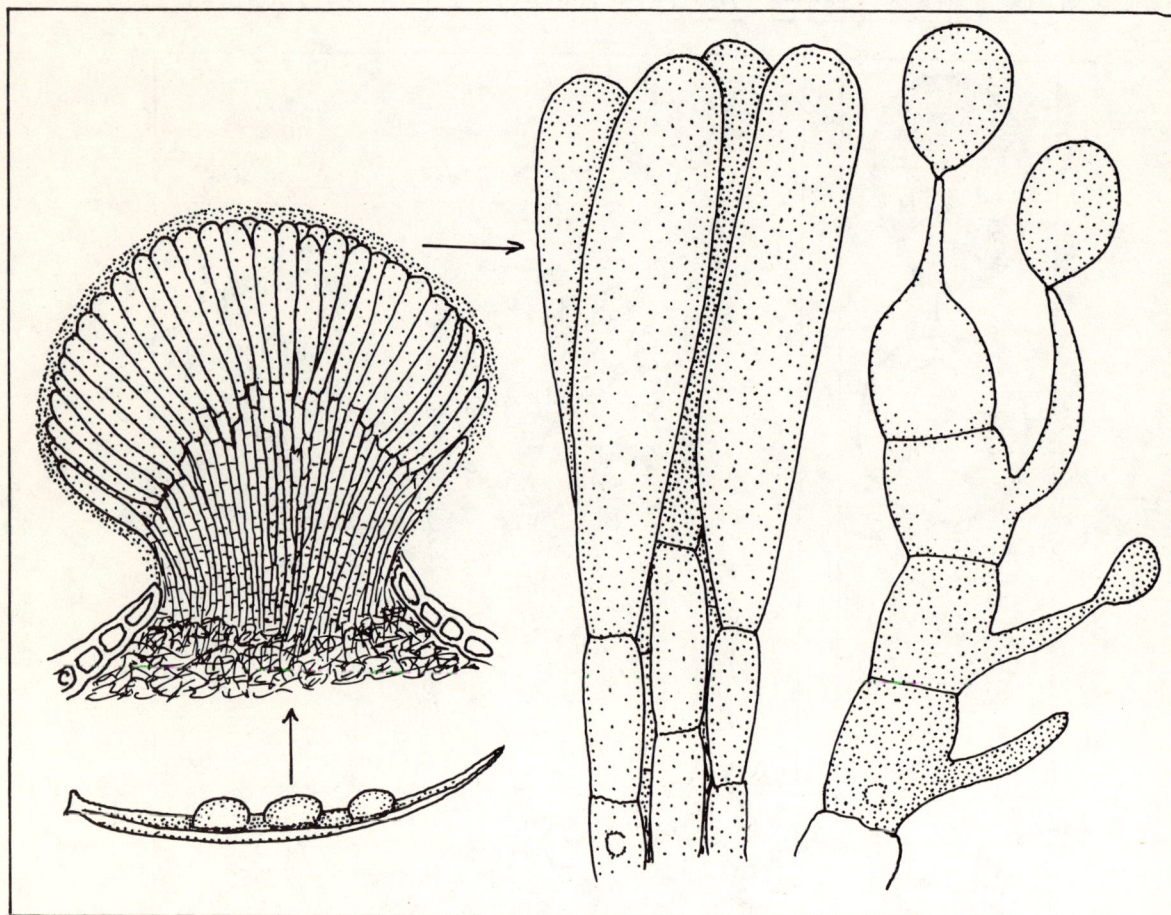

C. piceae (Barcl.) Bakshi & Singh; Telia (left) and teliospores.

44

17. CHRYSOCELIS Lagerheim & Dietel in Mayor, Mém. Soc. Neuch. Sci. Nat.
     5:542. 1913.

     Spermogonia subepidermal, type 4. Aecia subepidermal, without peridium,
opening by a pore-like rupture of the epidermis of the host, with or without
peridium; spores catenulate, verrucose. Uredinia subepidermal in origin, erumpent;
spores borne singly on pedicels, echinulate (or in C. lupini like the aecia and
aeciospores?), pores obscure. Telia subepidermal or substomatal in origin,
becoming erumpent, waxy when moist; spores 1-celled, laterally free, sessile, wall
thin and pale, germ pore not obviously differentiated, germination occurs by
elongation of the apex of the spore, basidium external, produced without dormancy.

     LECTOTYPE: Chrysocelis lupini Lagh. & Diet.

     Seven species have been recognized, two of which have spermogonia known. C.
lupini apparently has aecidioid uredinia, at least single sori occur that are
similar to the aecia except not accompanied by spermogonia. Host plants of the
species are not closely related, Acanthaceae, Cucurbitaceae, Leguminosae, Poly-
gonaceae, Rubiaceae, and Zingiberaceae. The genus differs from Chaconia mainly
because of type 4 spermogonia. C. geophilicola (Yen) comb. nov. (Stomatisora
geophilicola Yen) does not differ significantly. Hence we treat Stomatisora as
a synonym of Chrysocelis.

     REFERENCE: Ono, Y. 1978. Taxonomic revisions of the tribe Oliveae and
morphologically related genera (Uredinales). Ph.D. thesis Purdue University 254 pp.

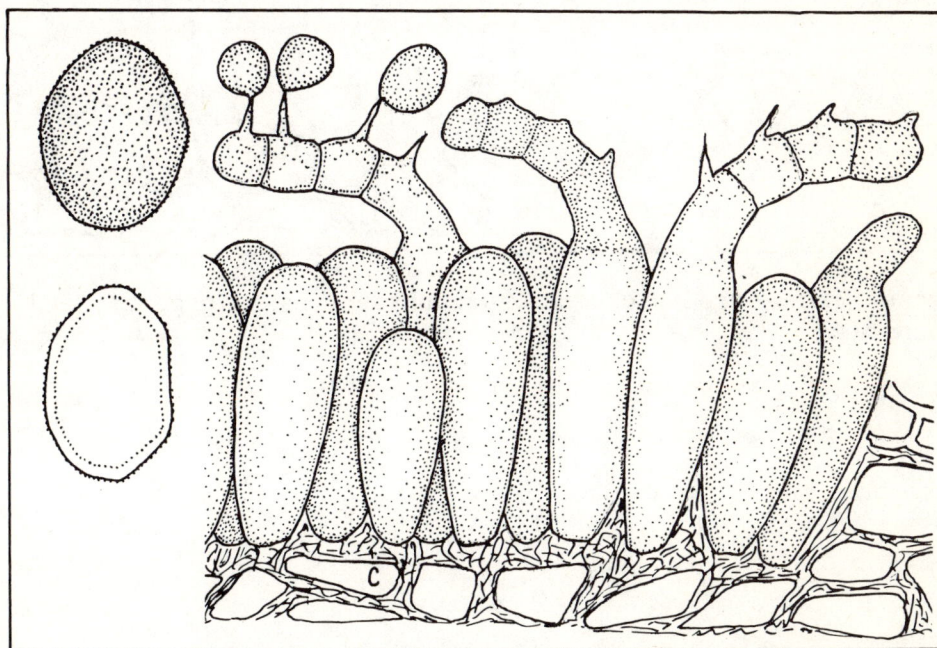

C. lupini Lagh. & Diet.; aeciospores and teliospores.

18. APLOPSORA Mains, Amer. J. Bot. 8:443. 1921.

Spermogonia and aecia unknown. Uredinia subepidermal in origin, only slightly erumpent, with peripheral paraphyses, united basally; spores borne singly on pedicels, pores obscure. Telia subepidermal in origin but soon exposed, consisting of a single layer of crowded, 1-celled, sessile, thin-walled, pale teliospores, germination occurs without dormancy, germ pore obscure if differentiated, basidium external.

TYPE: Aplopsora nyssae Mains.

Because the teliospores germinate without dormancy and the hosts are deciduous it is probable that the aecial state has perennial or at least over-wintering mycelium. The species probably are heteroecious. In addition to the type A. lonicerae Tranz. is known.
Aplopsora has uredinia similar to those of Phakopsora but the teliospores are borne singly. The paraphyses are unlike those of Melampsora. Ochropsora has internal basidia.

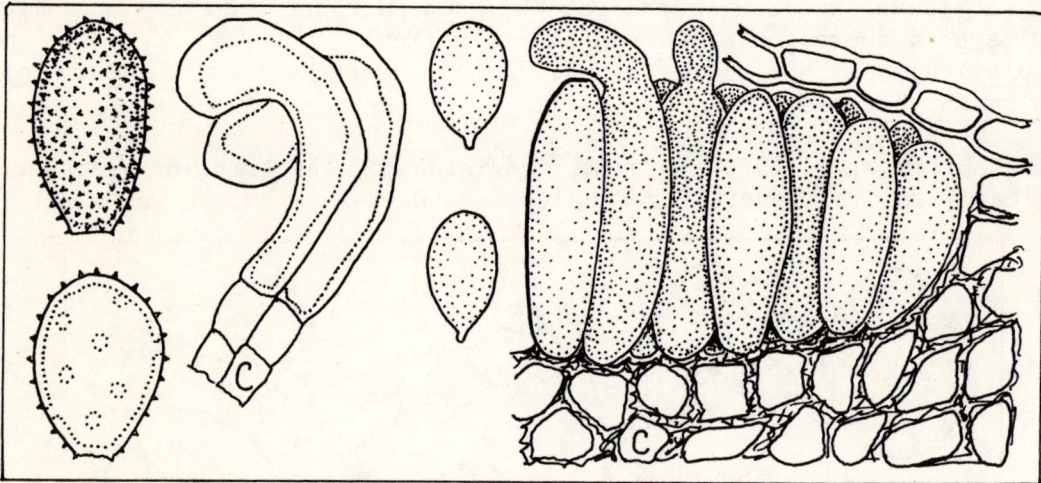

A. nyssae Mains; urediniospores, paraphyses, basidiospores, and teliospores.

19. CHACONIA Juel, Bihang. K. Svenska Vet. Akad. Handl. 23, Afd. 3, No. 10:12. 1897.

Spermogonia subcuticular, type 7, or subepidermal, type 5. Aecia subepidermal in origin, erumpent, uredinoid, paraphyses lacking or peripheral and incurved; spores borne singly on pedicels, echinulate, verrucose in lines, or more or less reticulate, pores equatorial or obscure. Uredinia similar to the aecia but not accompanied by spermogonia; spores like the aeciospores. Telia subepidermal in origin, erumpent; spores 1-celled, laterally free, sessile and grouped on sporogenous basal cells, germ pore probably not differentiated, wall thin and pale; germination occurs without dormancy; basidium external.

TYPE: Chaconia alutacea Juel.

Chaconia is a genus of about 6 species. The life cycles are known for half of the species and, where known, are macrocyclic and autoecious.

REFERENCES: Mains, E.B. 1938. Studies in the Uredinales, the genus Chaconia. Bull. Torrey Bot. Club 65:625-629. Ono, Y. 1978. Taxonomic revisions of the tribe Oliveae and morphologically related genera (Uredinales). Ph.D. thesis. Purdue University. 254 pp.

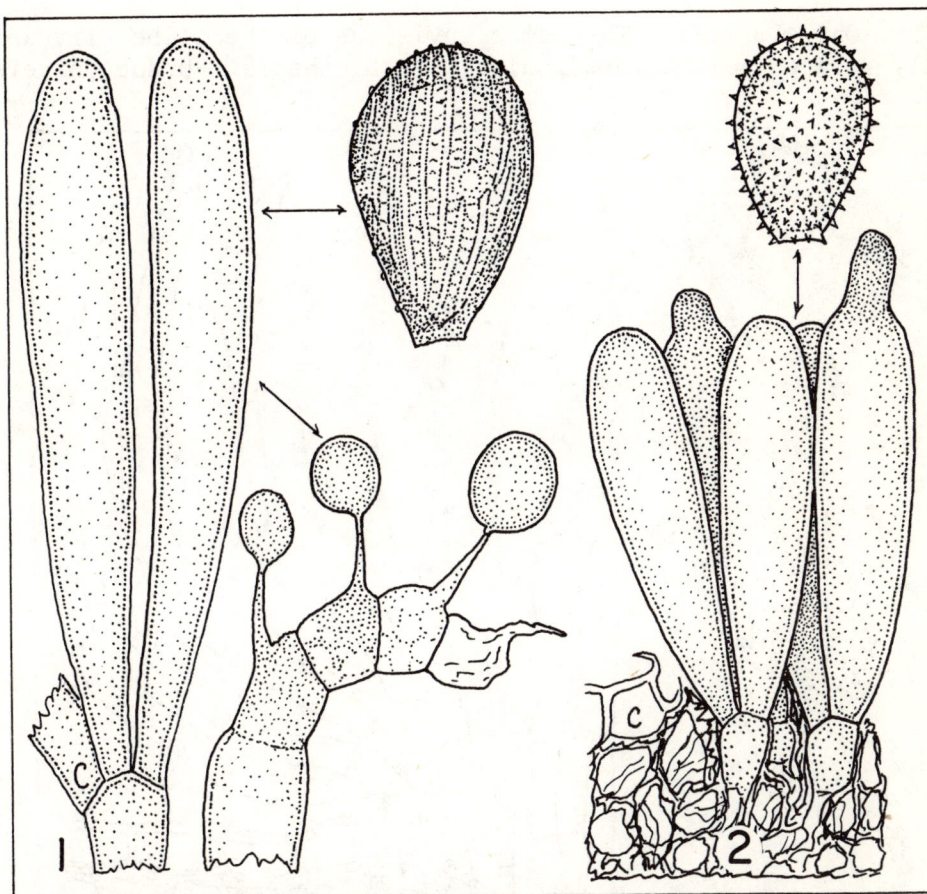

1. C. ingae (Syd.) Cumm.; Teliospores, a basidium, and an aeciospore. 2. C. alutacea Juel; one urediniospore and teliospores.

20. OLIVEA Arthur, Mycologia 9:60. 1917.

    Spermogonia subcuticular, type 7. Aecia subepidermal, deep-seated, opening by rupture of the epidermis, peridium lacking or subcuticular, erumpent, with peripheral paraphyses, uredinoid; spores catenulate in subepidermal aecia, verrucose with rod-like verrucae, or borne singly on pedicels in the uredinoid type and similar to urediniospores. Uredinia subepidermal or substomatal in origin, erumpent, with peripheral paraphyses united basally, spores borne singly on pedicels, echinulate, pores equatorial. Telia similar to uredinia in origin, erumpent; spores sessile on sporogenous basal cells, 1-celled, wall thin and pale; germ pore apparently not differentiated, germination occurs without dormancy, basidium external.

    TYPE: _Olivea capituliformis_ Arth.

    There are 7 or 8 species distributed in warm regions of the earth on members of the Euphorbiaceae, Labiatae, and Verbenaceae. Some species have strongly asymmetrical, lobed urediniospores or, when aecia are uredinoid, aeciospores. The latter is true of _Olivea fimbriata_ (Mains) comb. nov. (_Tegillum fimbriatum_ Mains, Bull. Torrey Bot. Club 67:707. 1940). We consider _Tegillum_ to represent a life-cycle variant rather than a distinctive genus.

    REFERENCE: Ono, Y. 1978. Taxonomic revisions of the tribe Oliveae and morphologically related genera (Uredinales). Ph.D. thesis. Purdue University, 254 pp.

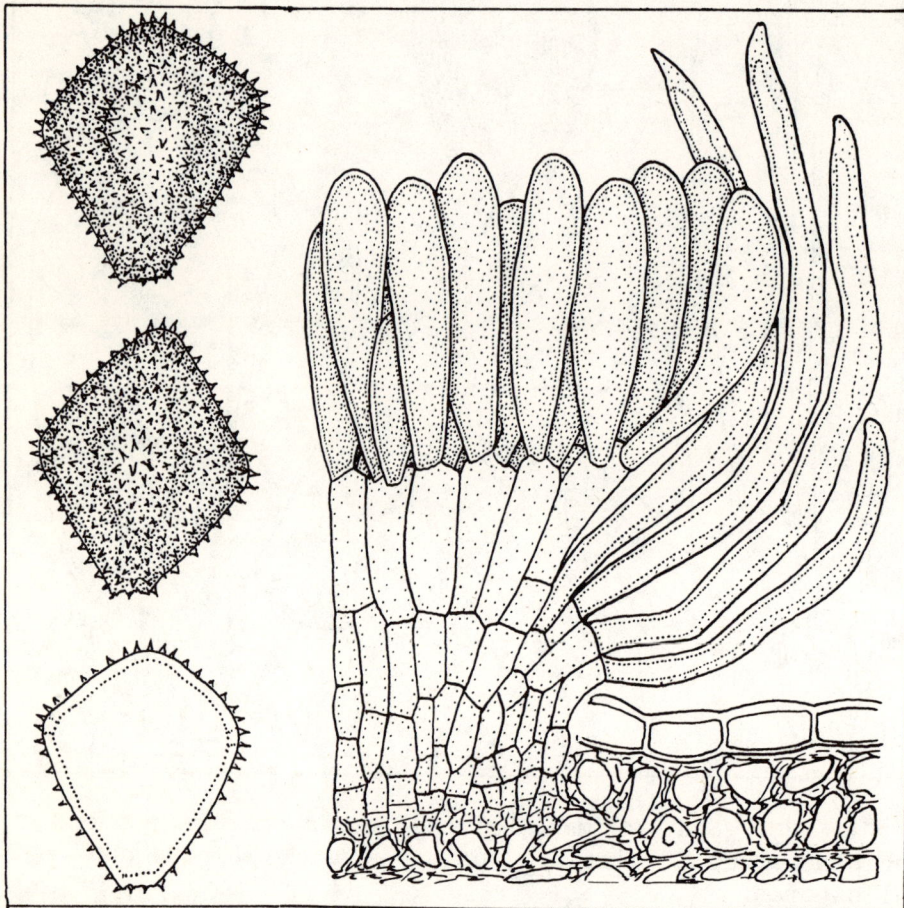

O. capituliformis Arth.; characteristically lobed urediniospores and paraphysate telia. The urediniospores of _O. fimbriata_ (Mains) Cumm. & Y. Hirat. are similar.

48

21. CHRYSOMYXA Unger, Beitr, Vergl. Pathol. p. 24. 1840.

Spermogonia subepidermal, type 2. Aecia subepidermal in origin, erumpent, with peridium, peridermioid; spores catenulate, verrucose. Uredinia subepidermal, erumpent, with or without an inconspicuous peridium; spores catenulate, verrucose, the urediniospores and aeciospores quite similar within a species, pores scattered. Telia subepidermal in origin, erumpent; spores 1-celled, catenulate, crowded but loosely adherent, wall thin, pale, germination occurs without dormancy, basidium external.

TYPE: Chrysomyxa abietis (Wallr.) Unger.

Except for the microcyclic species, C. abietis, C. weirii H.S. Jack., and C. deformans (Diet.) Jacz., the species are heteroecious with telia on Ericaceae, Empetraceae, and Aquifoliaceae and the aecia on needles and cones of Picea. C. arctostaphyli Diet. is demicyclic. No resting spores occur in the life cycles. Telia develop on overwintered, persistent leaves and the teliospores germinate without dormancy. Perennial witches' brooms on spruce bear the aecia of C. arctostaphyli and its telia are on Arctostaphylos uva-ursi. Aecia are not known for all species and others are assigned with doubt.

Of uncertain relationship is C.tsugae-unanensis Teng, a microcyclic species. Even more uncertain is the nature and relationship of Hiratsukaia tsugae (Hirat. f.) K. Hara (Chrysomyxa tsugae Hirat. f.). We exclude the genus.

REFERENCES: Savile, D.B.O. 1950. North American species of Chrysomyxa Can. J. Res. C. 28:318-330. Savile, D.B.O. 1955. Chrysomyxa in North America - additions and corrections. Can. J. Bot. 33:478-496. Ziller, W.G. 1974. The tree rusts of Western Canada. Ca. For. Serv. Publ. No. 1329. 272 pp. (Excellent for photographic illustrations.)

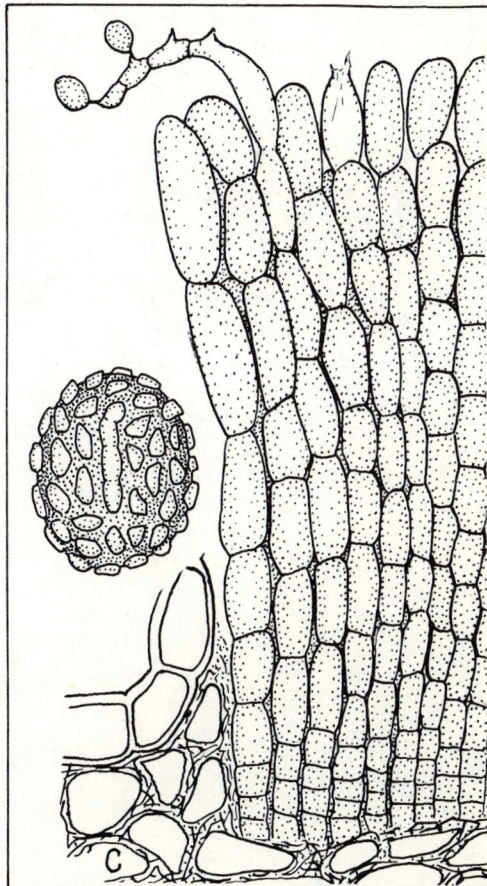

C. pirolata Wint.; urediniospore and telium.

22. ARTHURIA H.S. Jackson, Mycologia 23:463. 1931.

Spermogonia subcuticular, type 7. Aecia subepidermal in origin, becoming erumpent, caeomatoid; spores catenulate. Uredinia subepidermal, erumpent; spores
catenulate, echinulate, resembling the aeciospores, pores scattered. Telia subepidermal, erumpent; spores catenulate in loosely adherent chains, wall pale, germination occurs without dormancy, basidium external.

TYPE: Arthuria catenulata H.S. Jack. & Holw.

Three species have been described. All parasitize the Euphorbiaceae with two
species on Croton and one on Glochidion. In having all spores catenulate, Arthuria is similar to Chrysomyxa but differs because of echinulate rather than verrucose urediniospores and, more importantly, because Chrysomyxa has type 2 spermogonia. Patil and Thirumalachar (Indian Phytopathol. 23:610-614. 1970) proved the
life cycle of A. glochidionis Gokh., Patil & Thir. by inoculation.

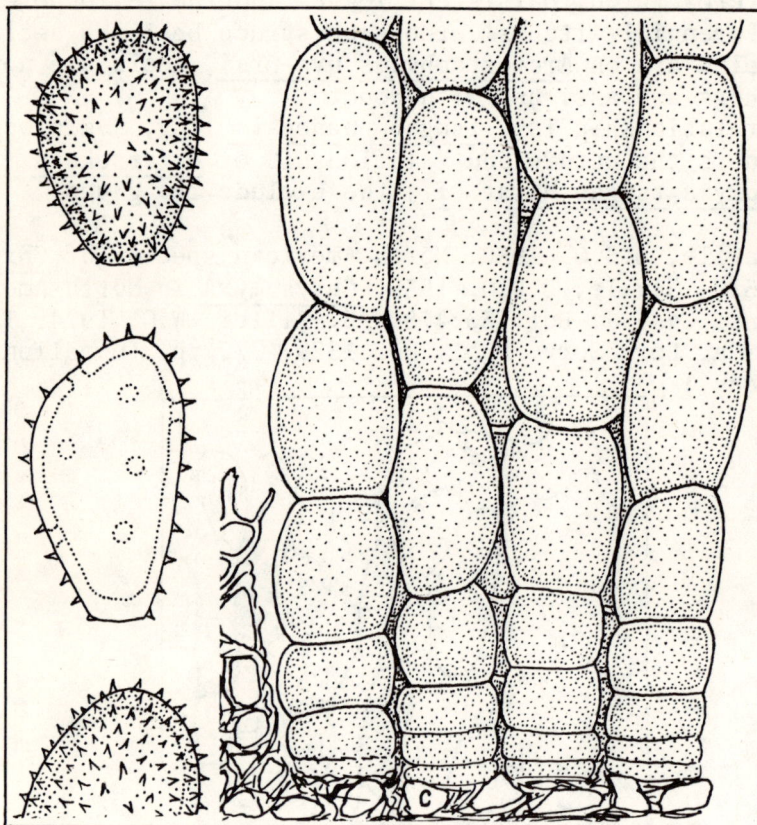

A catenulata; urediniospores (left). A. columbiana
(Kern & Whet.) Cumm.; a telium (right).

23. CEROTELIUM Arthur, Bull. Torrey Bot. Club 33:30. 1906.

Spermogonia subcuticular, type 7. Aecia subepidermal in origin, erumpent, with peridium, aecidioid; spores catenulate, verrucose. Uredinia subepidermal in origin, slightly erumpent, with peridium, peripheral, basally united paraphyses, or neither; spores borne singly, echinulate, pores scattered, obscure. Telia subepidermal in origin, becoming erumpent; spores catenulate in short chains, scarcely adherent laterally, wall thin and pale, germ pore obscure if differentiated, germination occurs without dormancy, basidium external.

TYPE: Cerotelium canavaliae Arth.

Aecia are known only for C. dicentrae Mains & H.W. Ander., which produces aecia on Dicentra cucullaria and telia on Laportea canadensis. The aecial state is systemic in C. dicentrae and, because teliospores germinate without dormancy, it is probable that a persistent aecial state of some sort may occur in most or all species. The uredinia of C. canavaliae have a peridium, of C. fici (Butl.) Arth. peripheral paraphyses. The number of species is uncertain but probably not large. C. fici is the commonest species and is circumglobal in warm regions.

REFERENCES: Laundon, G.F. 1975. Taxonomy and nomenclature notes on Uredinales. Mycotaxon 3:133-161. Thirumalachar M.J. 1960. Critical notes on some plant rusts III. Mycologia 52:688-693. Both with reference to Cerotelium, Catenulopsora, and Kuehneola.

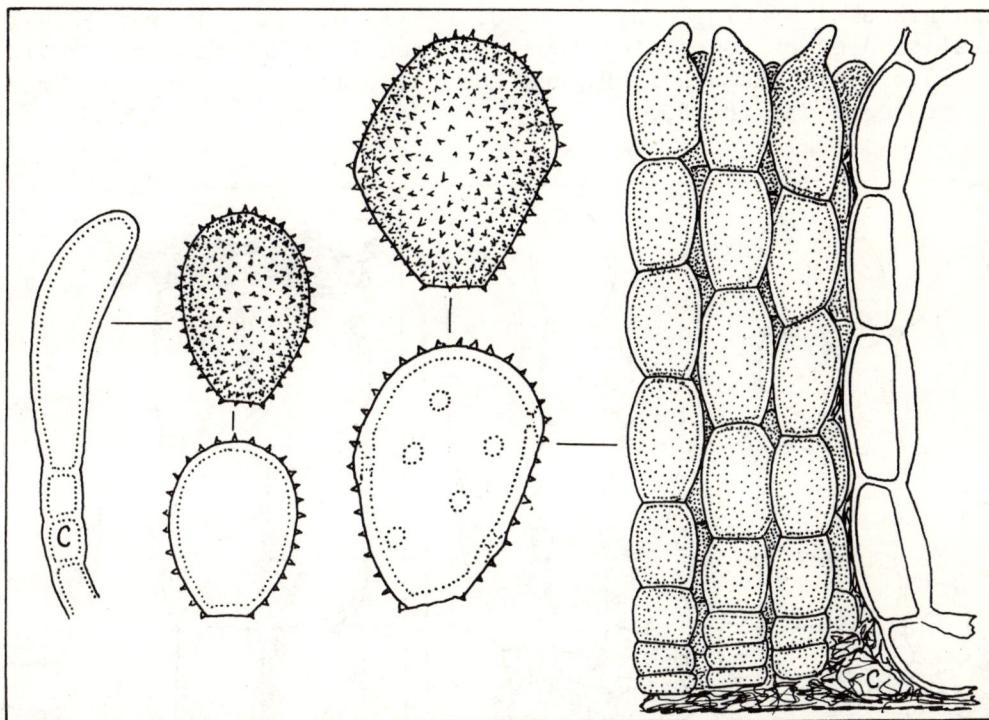

C. fici (Butl.) Arth.; a paraphysis and two urediniospores (left).
C. canavaliae Arth.; two urediniospores and a telium (right).

24. PHRAGMIDIELLA P. Hennings, Bot. Jahrb. 38:104. 1905.

Spermogonia subcuticular, type 7. Aecia subcuticular in origin, erumpent, uredinoid; spores borne singly, echinulate. Uredinia subepidermal in origin, erumpent; spores borne singly, echinulate, pores equatorial. Telia subepidermal in origin, mostly erumpent, composed of crowded but not adherent chains of 1-celled spores, germ pore 1, apical and indistinct, wall pale, germination occurs without dormancy, basidium external.

TYPE: Phragmidiella markhamiae P. Henn.

There are four or five species in Africa and India. Most parasitize plants of the Bignoniaceae. Insofar as the life cycles are known, the species are autoecious and macrocyclic. The genus has telia similar to Cerotelium; both genera have type 7 spermogonia; Cerotelium has aecidioid aecia but the aecia are uredinoid in Phragmidiella; and the urediniospores are quite unlike. It is doubtful that the genera are closely related. Patil and Thirumalachar proved the life cycle for P. stereospermi (Mund.) Thir. & Mund. by inoculation. Mundkur and Thirumalachar described the development of P. heterophragmae. Mehtamyces Mund. & Thir. (M. stereospermi, type) is treated here as a synonym differs because the telia are not erumpent but the urediniospores are much as those of P. heterophragmae and the life cycle is as other species of Phragmidiella. Ramachar and Rao have proposed reinstatement of Mehtamyces.

REFERENCE: Mundkur, B.B. and Thirumalachar, M.J. 1945. Two new genera of rusts on Bignoniaceae. Mycologia 37:619-628. Patil, B.V. and Thirumalachar, M.J. 1970. Cultural studies on some autoecious rust fungi in Maharashtra. Indian Phytopathol. 23:610-614. Ramachar, P. and Rao, A.S. 1981. Mehtamyces (Uredinales) reinstated. Mycologia 73:778-780.

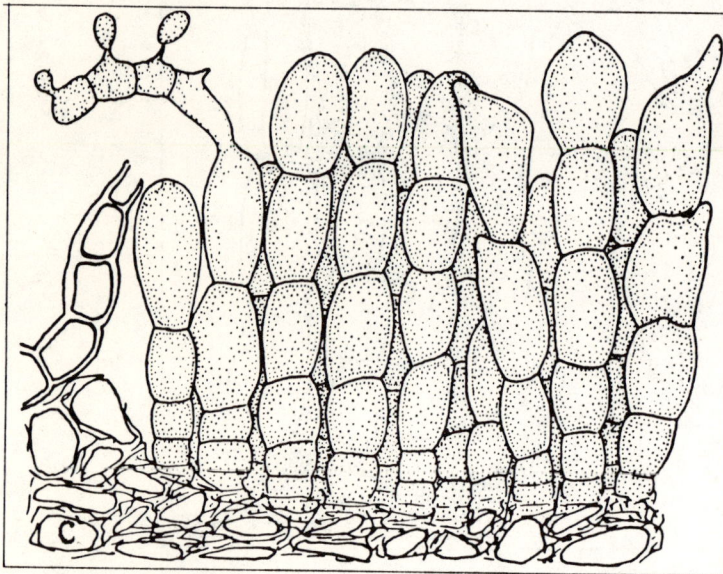

P. markhamiae P. Henn.; telium (above). P. stereospermi (Mundk.); teliospores and urediniospores (right).

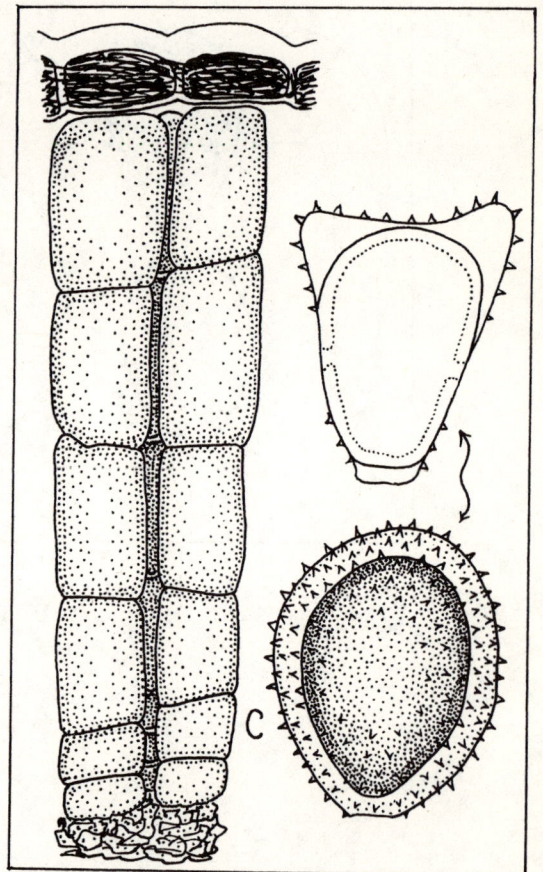

25. KWEILINGIA Teng, Sinensia 11:124. 1940.

Spermogonia and aecia unknown. Uredinia subepidermal in origin, erumpent; spores borne singly, echinulate, pores obscure. Telia subepidermal in origin, erumpent and spreading over the leaf and sheath as a mantle; spores 1-celled, catenulate or perhaps on pedicels and multicellular, the spores or chains loosely adherent at first but separating later, wall brown, germ pores 2, apical or nearly in each cell, basidium doubtless external.

TYPE: Kweilingia bambusae Teng.

The type was described from Kwangsi, China on Bambusa sp. and has not been reported elsewhere unless, as seems probable, Tunicopsora bagchii Singh & Pand. is synonymous. Certainly Tunicopsora is synonymous with Kweilingia. T bagchii occurs on Dendrocalamus strictus in India. Contaminant fungi colonize the sori early and make accurate observation difficult. Thirumalachar and Narasimhan decided that the fungus is not a rust but a member of the Auriculariales, based on a study of the type of Kweilingia.

REFERENCES: Singh, S. and Pandy, P.C. 1971. Tunicopsora, a new rust genus on bamboo. Trans. Brit. Mycol. Soc. 56:301-318. Thirumalachar, M.J. and Narasimhan, M.J. 1951. Critical notes on some plant rusts. III. Sydowia 5:476-483.

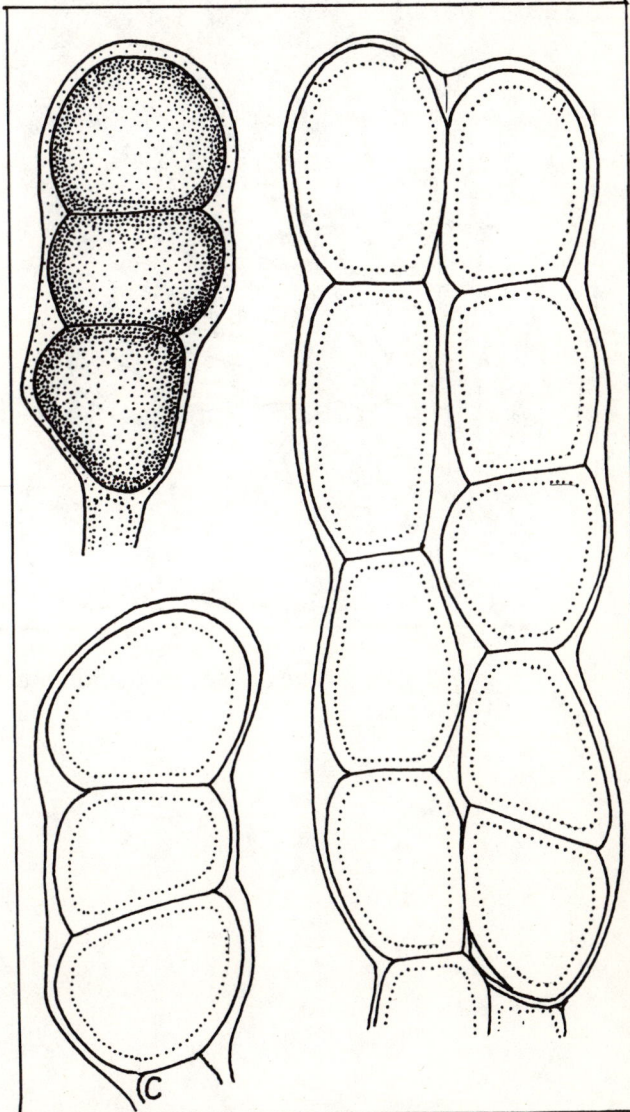

K. bambusae Teng; teliospores.

26.   UREDOPELTIS P. Hennings, Ann. Mus. Congo Belge 2:223. 1908.

Spermogonia and aecia unknown. Uredinia subepidermal in origin, becoming erumpent, small, with incurved, basally united, peripheral paraphyses; spores borne singly, echinulate, pores scattered. Telia subepidermal in origin, becoming strongly erumpent, consisting of cushions of laterally adherent spores several spores deep; spores sessile and irregularly arranged, not catenulate, 1-celled, germ pore probably 1, basidium not seen but doubtless external.

TYPE: Uredopeltis congensis P. Henn.

Uredopeltis appears to be little more than an erumpent Phakopsora. Laundon (Brit. Mycol. Soc. Trans. 46:503-504. 1963) studied the type specimen from the Congo and specimens from Northern Rhodesia. In the 1959 "Illustrated Genera" Cummins erroneously treated Uredopeltis as a synonym of Phragmidiella. Both parasitize Markhamia of the Bignoniaceae. U. congensis is the only species known.

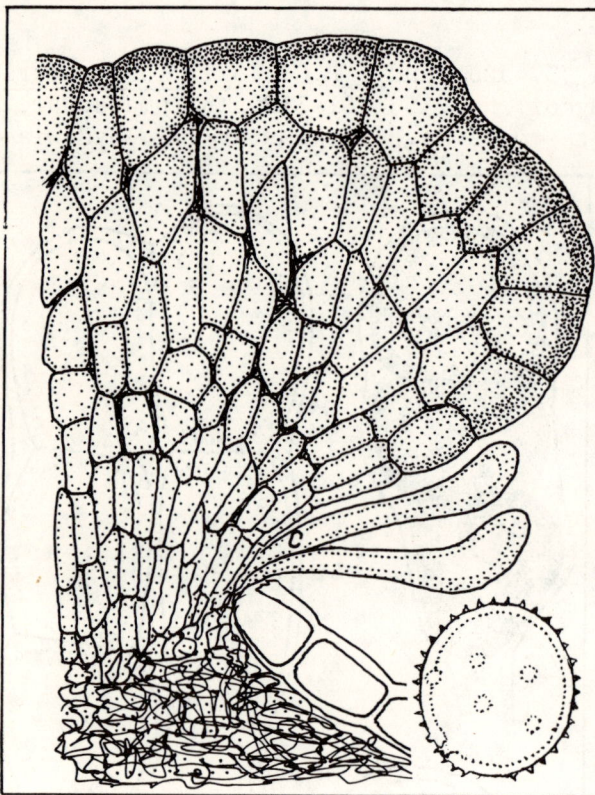

U. congensis P. Henn.; telium and one urediniospore.

27. DASTURELLA Mundkur & Kheswalla, Mycologia 35:202. 1943.

Spermogonia subcuticular, type 7. Aecia subepidermal in origin, erumpent with peridium, aecidioid; spores catenulate, verrucose. Uredinia subepidermal, erumpent, with abundant, incurved, peripheral paraphyses; spores borne singly, echinulate, pores equatorial. Telia subepidermal in origin, becoming erumpent in cushion-like crusts several spores deep; spores 1-celled, firmly adherent laterally and terminally, catenulate, brown, germ pore not seen, basidium external.

TYPE: Dasturella divina (Syd.) Mund. & Khes. (Angiopsora divina Syd.).

Only two or three species are known. The type species is heteroecious with aecia on Randia dumetorum (Rubiaceae) and the telia on Dendrocalamus (Gramineae). Dasturella differs from Physopella because of the strongly erumpent telia and from Uredopeltis because of catenulate teliospores.

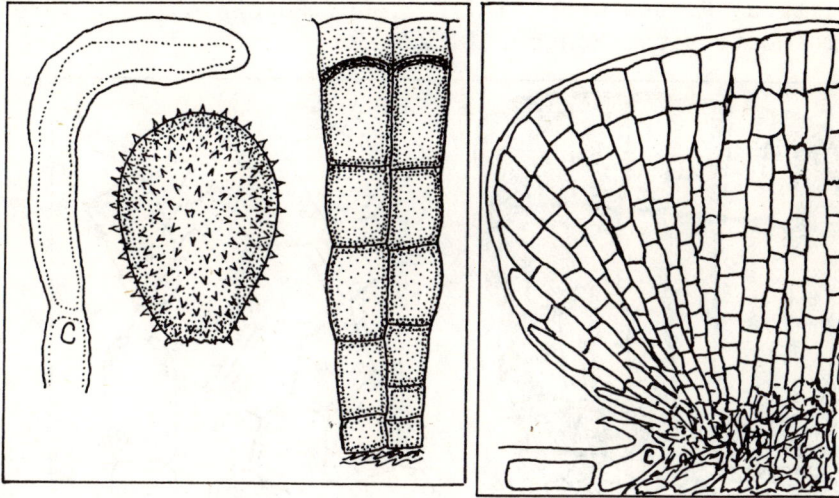

D. divina Mundk. & Khes.; one paraphysis, one uredini-iospore, teliospore chains, and a telium.

28.   ALVEOLARIA Lagerheim, Ber. Dtsch. Bot. Ges. 9:346-347. 1891 (issued 1892).

Spermogonia subepidermal, type 4. Aecia and uredinia not produced. Telia subepidermal, deep-seated, becoming erumpent as short columns of strongly adherent radial plates, the plates rather easily separable; spores 1-celled, germ pore 1, apical, germination occurs without dormancy, basidium external.

TYPE: Alveolaria cordiae Lagh.

Only the type and A. andina are recognized; both parasitize species of Cordia (Boraginaceae). The arrangement of single layers of teliospores in plates is unique and gives no clue to the relationship of the genus. In A.cordiae the telia are surrounded by very long hypertrophied cells of the mesophyll.
Buriticá and Hennen have described both subepidermal and subcuticular spermogonia but we question the nature of the subcuticular structures.

REFERENCE: Buriticá, P. and Hennen, J.F. 1980. Pucciniosireae (Uredinales - Pucciniaceae). Flora Neotropica, Monogr. 24:22-23.

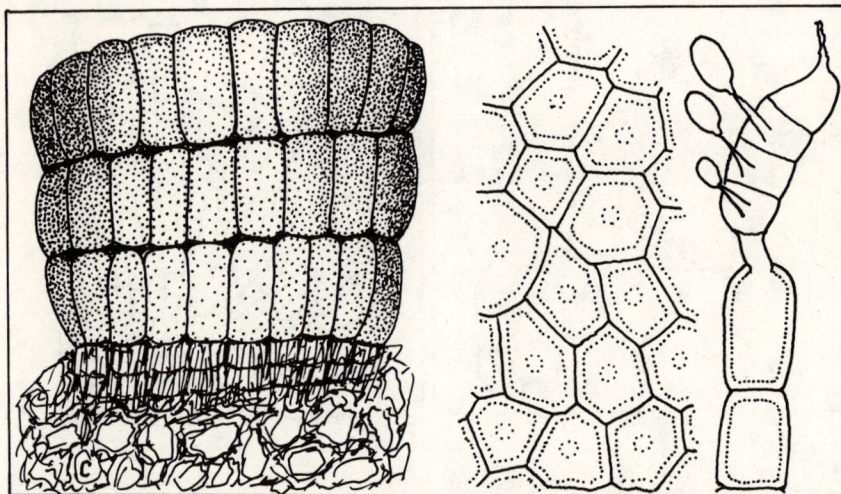

A. cordiae Lagh.; telium, surface view of a plate of teliospores, and a germinating teliospore.

29. BAEODROMUS Arthur, Ann. Mycol. 3:19. 1905.

Spermogonia subepidermal, type 4. Aecia and uredinia not produced. Telia subepidermal in origin, erumpent, compact; spores 1-celled, catenulate in adherent chains of few spores each or irregularly arranged, wall mostly lightly pigmented, germpore obscure, probably apical, perhaps not differentiated, basidium external.

LECTOTYPE: Baeodromus holwayi Arth.

Only seven species have been described and all are microcyclic, occurring on Senecio (five species), Croton (one species), and Urtica (one species). The relationship of the genus is obscure. Azbukina established a tribe, Baeodromeae, for this genus.

REFERENCES: Azbukina, Z.M. 1974. Rust fungi of the Soviet Far East. (in Russian). Akad. Nauk, Moscow. 527 pp. (p.149). Buriticá, P. and Hennen, J.F. 1980. Pucciniosireae (Uredinales - Pucciniaceae). Flora Neotropica, Monogr. 24:23-28.

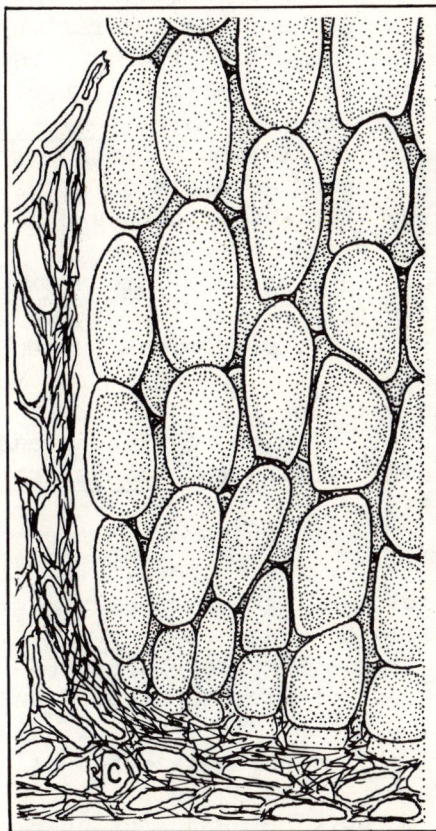

B. eupatorii (Arth.) Arth.;
telium.

30. CRONARTIUM Fries, Obs. Mycol. 1:220. 1815.

Spermogonia intracortical, under the periderm, type 9. Aecia intracortical in origin, erumpent, most with strongly developed peridium, peridermioid; infections perennial and commonly causing hypertropy of stems or cones; spores catenulate, verrucose with rod-like columns mostly having annulate sides. Uredinia subepidermal in origin, opening by a pore, with peridium and sometimes also with intrasoral paraphyses; spores borne singly, echinulate, germ pores scattered or bizonate, obscure. Telia subepidermal in origin, often arising in uredinia, becoming erumpent as hair-like columns of strongly adherent spores embedded in a common matrix; spores 1-celled, catenulate, pores 1-3, obscure, wall pale, germination occurs without dormancy, basidium external.

TYPE: Cronartium flaccidum (Alb. & Schw.) Wint. (Cronartium asclepiadeum Fries).

The species of Cronartium are heteroecious, producing spermogonia and aecia on the stems or cones of Pinus spp. and the uredinia and telia on various dicotyledonous plants. Because the teliospores are not resting spores, a persistent aecial state is to be expected and occurs, either in stems or cones. The period between infection of the pine and the production of aecia varies from two to three to many years. When cones are infected, as in C. conigenum Hedgc. & Hunt and C. strobilinum Hedgc. & Hahn, infections terminate with sporulation because the affected cone dies. In species that produce aecia on stems and branches, e.g. C. ribicola J.C. Fisch., C. quercuum (Berk.) Miyabe, and C. coleosporioides Arth., the aecial cankers may sporulate for years.

The economic importance of the species of Cronartium is determined partly by the value of the pine host and partly by the type of injury inflicted. In northern areas where white pine are important timber trees, white pine blister rust (caused by C. ribicola) is of importance; in the southern states of the U.S. the fusiform gall rust (caused by C. quercuum f. sp. fusiforme) is the important species.

There are authenticated cases of pine-to-pine spread. Some (Peridermium harknessii and P. pini have been referred to Endocronartium (which see) and perhaps others will be.

REFERENCES: Antonopoulos, A.A. and Chapman, R.L. 1976. Morphology of Cronartium fusiforme aeciospores: A light and scanning electron microscopic study. Bot. Gaz. 137:285-289. Burdsall, H.H. Jr. and Snow, G.A. 1977. Taxonomy of Cronartium quercuum and C. fusiforme. Mycologia 69:503-508. Grand, L.F. and Moore, R.T. 1972. Scanning electron microscopy of Cronartium spores. Can. J. Bot. 50:1741-1742. Henderson, D.M. and Hiratsuka, Y. 1974. Ontogeny of spore markings on aeciospores of Cronartium comandrae and peridermioid teliospores of Endocronartium harknessii. Can. J. Bot. 52:1919-1921. Hiratsuka, Y. 1971. Spore surface morphology of pine stem rusts of Canada as observed under a scanning electron microscope. Can. J. Bot. 49:371-372. Hiratsuka, Y. and Powell, J.M. 1976. Pine stem rusts of Canada. Can. For. Serv. For. Tech. Rept. No. 4. 82 pp. Peterson, R.S. 1967. The Peridermium species on pine stems. Bull. Torrey Bot. Club 94:511-542. Peterson, R.S. 1973. Studies of Cronartium (Uredinales). Rept. Tottori Mycol. Inst. Japan 10:203-223.

Illustrations next page.

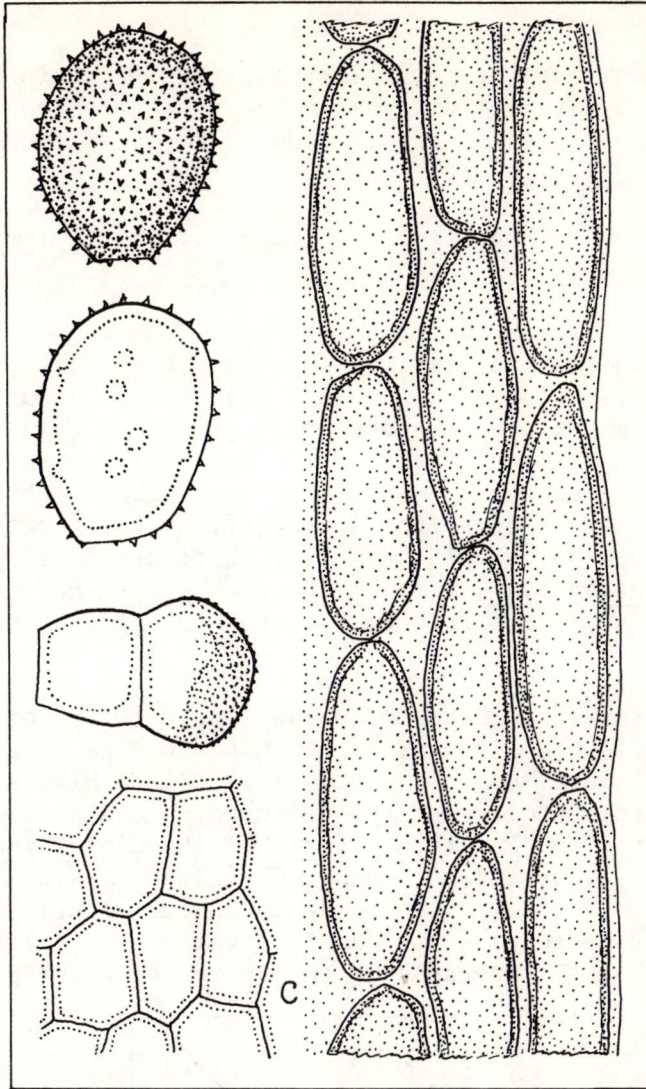

C. *comandrae* Peck; urediniospores, ostiolar cell of the peridium, side cells of
peridium in surface view, and teliospores
and the matrix.

30E. ENDOCRONARTIUM Y. Hiratsuka, Can. J. Bot. 47:1493. 1969.

Spermogonia intracortical, type 9. Telia intracortical in origin, becoming erumpent, indeterminate, with peridium as in the aecia of Cronartium; spores 1-celled, catenulate, verrucose with tapered, columnar rods, wall colorless, basidium external.

TYPE: Endocronartium harknessii (J.P. Moore) Y. Hirat. (Peridermium harknessii (J.P. Moore).

E. harknessii, the western gall rust, occurs in North America; E. pini occurs in Europe. Endocronartium can be distinguished from the aecial state of Cronartium only by germinating the spores and hence is a life cycle "genus" comparable to Endophyllum vs. Aecidium. Hiratsuka showed that spores of Peridermium harknessii and P pini produce a basidium in which nuclear fusion followed by meiosis occur, but instead of basidiospores there result short, uninucleate branches which presumably function as infection pegs. Laundon questions whether there is proof that nuclear fusion and meiosis occur. In any case, there are four species known to infect pine-to-pine, P. harknessii, P. pini, P. yamabense Saho., and two races of P. filamentosum Peck. Only the first two have been treated as species of Endocronartium.

REFERENCES: Hiratsuka, Y. 1968. Morphology and cytology of aeciospores and aeciospore germ tubes of host-alternating and pine-to-pine races of Cronartium flaccidum in Northern Europe. Can. J. Bot. 46:1119-1122. Hiratsuka, Y., Morph, W. and Powell, J.M. 1966. Cytology of the aeciospores and aeciospore germ tubes of Peridermium harknessii and P. stalactiforme of the Cronartium coleosporioides complex. Can. J. Bot. 44:1639-1643. Laundon, G.F. 1976. PERIDERMIUM (Fungi). Taxon 25:186-187. Saho, H. 1981. Notes on Japanese rust fungi VII Peridermium yamabense sp. nov., a pine-to-pine stem rust of white pines. Trans. Mycol. Soc. Japan 22:27-36. (incl. SEM micrographs of the spores of three pine-to-pine species.)

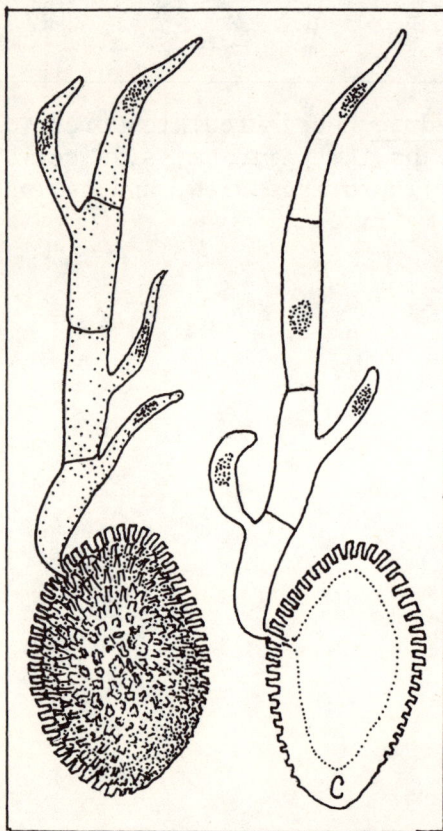

E. harknessii; teliospores.

31. CROSSOPSORA H. & P. Sydow, Ann. Mycol. 16:243. 1919.

Spermogonia subcuticular, type 7. Aecia subepidermal, opening by a pore-like rupture of the epidermis, without a peridium; spores catenulate, echinulate. Uredinia subepidermal in origin, becoming erumpent, with usually septate and basally united paraphyses; spores borne singly, echinulate, pores scattered, obscure. Telia subepidermal in origin, erumpent as hair-like columns of 1-celled, catenulate, strongly adherent spores, germ pores not observed, germination occurs without dormancy, basidium external.

TYPE: Crossopsora zizyphi (Syd. & Butl.) Syd. (Cronartium zizyphi Syd. & Butl.).

The genus has not been studied as a unit but there are 12 to 15 species and all inhabit warm regions. There is a superficial resemblance of the telia to those of Cronartium but it is doubtful that the two genera are related. The spermogonia are unlike, the aecia are not peridermioid, the uredinia have paraphyses rather than a peridium, and the teliospores are not embedded in a common matrix. Mundkur and Thirumalachar describe and illustrate superstomatal uredinia in C. zizyphi but whether most species have such sori is doubtful.

No general statement can be made about life histories because spermogonia and aecia have been described only in C. sawadae (H. & P. Syd.) Arth. & Cumm. But it is probable that all species will prove to be autoecious.

REFERENCES: Arthur, J.C. and Cummins, G.B. 1936. Philippine rusts in the Clemens Collection 1923-1926, II. Phil. J. Sci. 61:463-488. Peterson, R.S. 1973. Studies of Cronartium (Uredinales). Rept. Tottori Mycol. Inst. (Japan). No. 10: 203-223. Mundkur, B.B. and Thirumalachar, M.J. 1946. Revisions of and additions to Indian Fungi I. Commonw. Mycol. Inst. Mycol. Pap. No. 16:1-27.

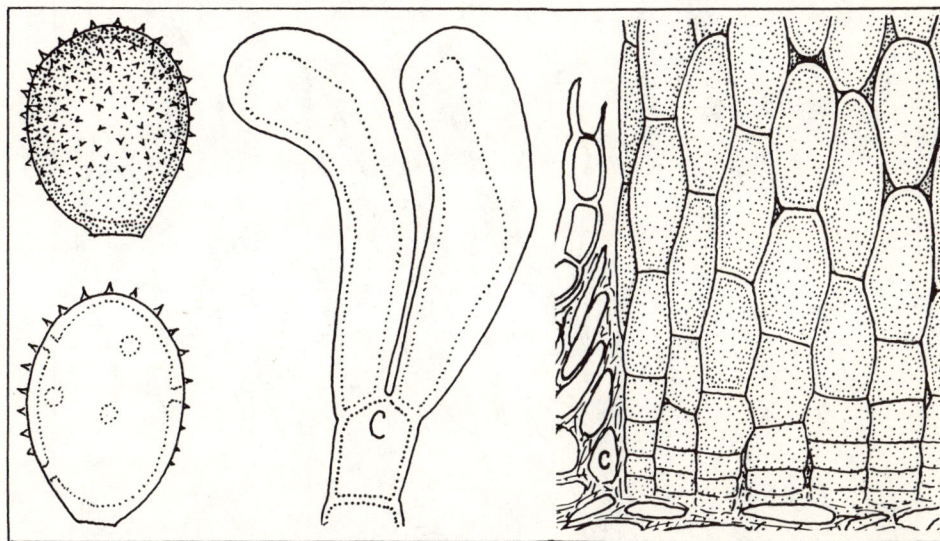

C. sawadae (Syd.) Arth. & Cumm.; urediniospores and paraphyses (left). C. zizyphi (Syd. & Butl.) Syd.; telium (right).

61

32. SKIERKA Raciborski, Paras. Algen Pilze Javas. 2:30. 1900.

Spermogonia subepidermal, type 5. Aecia subepidermal in origin, opening by a pore-like rupture of the epidermis, without peridium, uredinoid; spores borne singly. Uredinia subepidermal, similar to the aecia except not associated with spermogonia; spores similar to the aeciospores, the wall thickened in two opposite, longitudinal, somewhat hygroscopic bands or ridges, the crest of which may be echinulate or crenulate. Telia subepidermal in origin, as the uredinia; spores 1-celled, sessile, produced in irregular succession, strongly adherent and extruded in long, hair-like columns, germ pore not obvious but doubtless 1 per cell, germination occurs without dormancy, basidium external.

TYPE: Skierka canarii Racib.

There are some 10 species distributed circumglobally in the tropics. All are autoecious and presumably macrocyclic but only telia have been described for S. robusta Doidge. The species occur about equally on Geraniales and Sapindales and with one species on Rhamnales. Skierka differs form Cionothrix in the type of spermogonium.

REFERENCE: Mains, E.B. 1939. The genera Skierka and Ctenoderma. Mycologia 31:175-190.

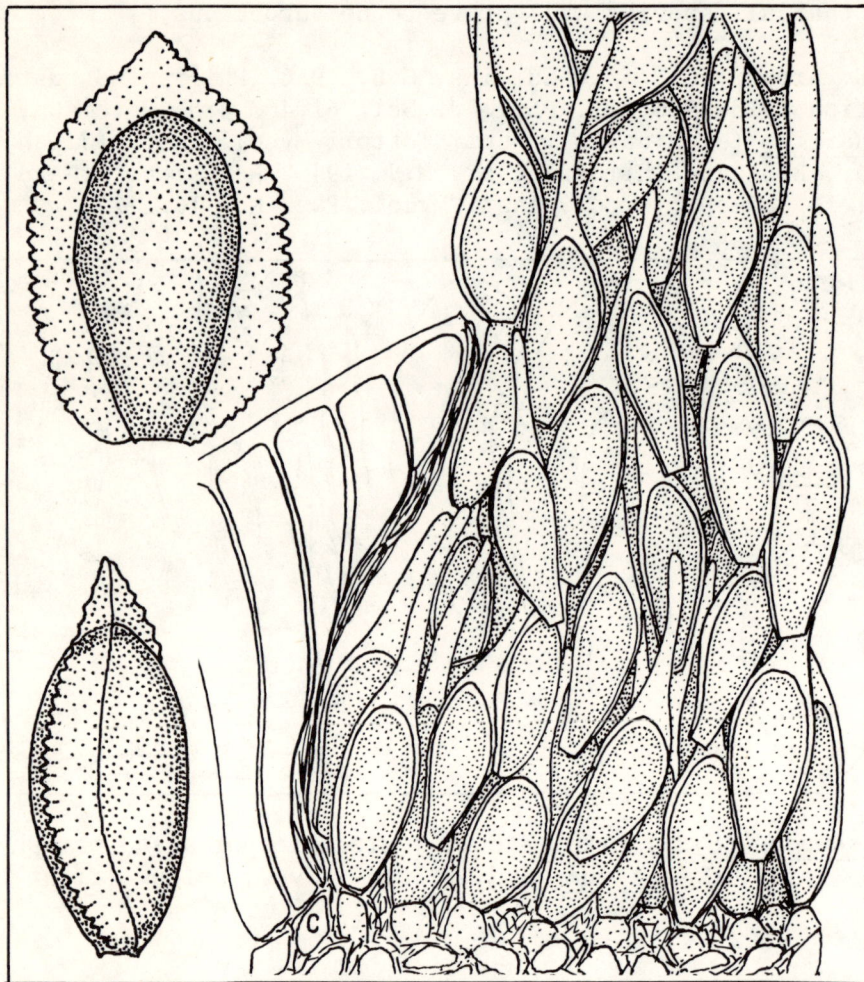

S. holwayi Arth.; urediniospores and a telium.

33. MASSEEËLLA Dietel, Ber. Dtsch. Bot. Ges. 13:332. 1895.

Spermogonia subcuticular, type 7. Aecia subepidermal in origin, becoming erumpent, aecidioid with peridium or caeomatoid without peridium; spores catenulate, verrucose. Uredinia subepidermal in origin, small, slightly erumpent; spores borne singly, echinulate, pores obscure. Telia subepidermal in origin, deep-seated, lined with upwardly directed hyphae, which may secrete the gelatinous matrix in which the spores are embedded, and extruded in hair-like columns; spores 1-celled, sessile, without order in the matrix, not adherent, wall pigmented, germ pore 1, basidium external.

TYPE: Masseeëlla capparidis (Hobs.) Diet. (Cronartium capparidis Hobs.).

All species occur in warm regions from India to the Philippines. Insofar as known, all species are macrocyclic and autoecious. M. narasimhanii Thir. is the only species with caeomatoid aecia reported. This difference was used by Sathe as the basis for the genus Kamatomyces which we treat as a synonym. It is not unusual for a genus to have different kinds of aecia. Thirumalachar, when he described M. narasimhanii, commented "In respect to the structure and development of the teliospores all the species of Masseeëlla so far known, show remarkable resemblance."

REFERENCES: Sathe, A.V. 1965. Revision of Masseeella narasimhanii Thirum. (Uredinales). Sydowia 19:187-189. Thirumalachar, M.J. 1943. Masseeëlla narasimhanii, a new species of rust on Flueggia leucocarpus Willd. Indian Acad. Sci. Proc. 18B:36-40.

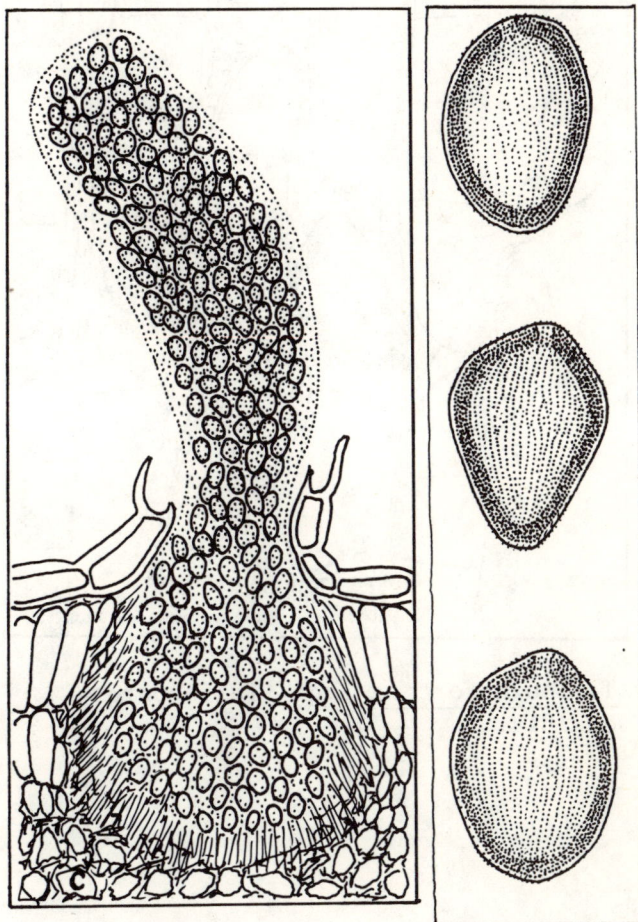

M. flueggiae Syd.; telium and teliospores.

34. TRICHOPSORA Lagerheim, Ber. Dtsch. Bot. Ges. 9:347. 1891 (issued 1892).

Spermogonia subepidermal, type 4. Aecia and uredinia not produced. Basidiosori subepidermal in origin, becoming erumpent as more or less gelatinous, hair-
like columns; basidia 1-celled, catenulate, with elongating and gelatinizing intercallary cells, becoming basidia by division of the protoplast into 4 cells each
of which produces 1 basidiospore; germination occurs without dormancy.

TYPE: Trichopsora tournefortiae Lagh.

Only one species is known and it occurs in South America on Tournefortia of
the Boraginaceae. The basidial column has a superficial resemblance to that of
Cronartium but the germination is different. Coleosporium crowellii Cumm. lacks
the intercallary cells but otherwise is similar in appearance. The relationship
of the genus is obscure, but Buriticá and Hennen indicate relation to Chardoniella.

REFERENCE: Buriticá, P. and Hennen J.F. 1980. Pucciniosireae (Uredinales)-
Pucciniaceae). Flora Neotropica, Monogr. 24:42-44.

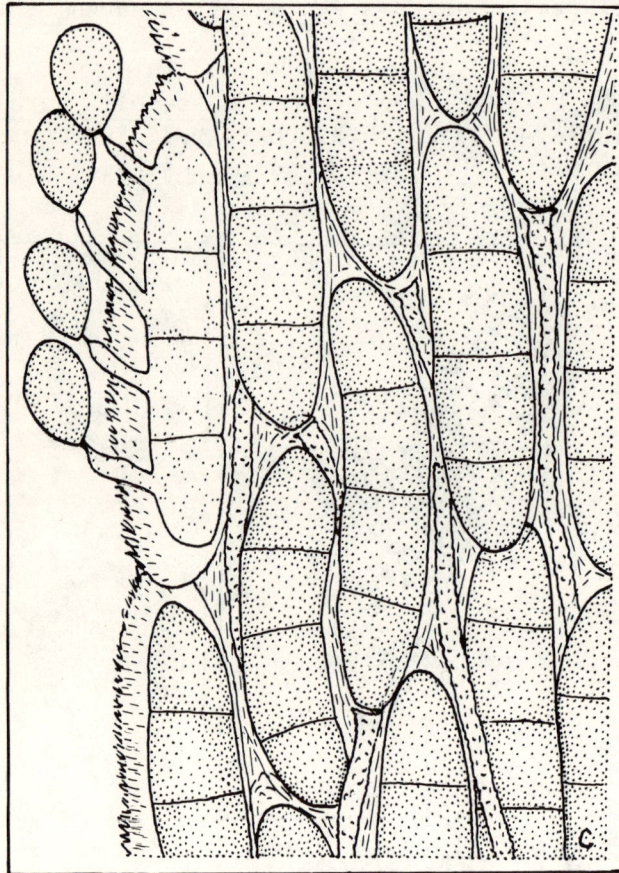

T. tournefortiae Lagh. Teliospores=
basidia.

35. CIONOTHRIX Arthur, N. Amer. Flora 7:124. 1907.

Spermogonia subepidermal, type 4. Aecia and uredinia unknown, probably not produced. Telia subepidermal in origin, with peripheral paraphyses; spores catenulate in origin but soon irregularly arranged and extruded in long flexuous, hair-like columns of adherent spores, spores 1-celled, germ pore perhaps not differentiated, germination occurs without dormancy by elongation of the apex of the spore to form the basidium, basidium external.

TYPE: Cionothrix praelonga (Wint.)Arth. (Cronartium praelongum Wint.).

Cionothrix is generally similar to Skierka except for some differences in the form of the teliospores and the type of spermogonium. Some species, including the type, were originally described as Cronartium and considered to be microcyclic species of that genus. All species are microcyclic. The relationships of the genus are obscure. A few species parasitize the Compositae.

REFERENCE: Buriticá, P. and Hennen, J.F. 1980. Pucciniosireae (Uredinales - Pucciniaceae). Flora Neotropica, Monogr. 24:18-22.

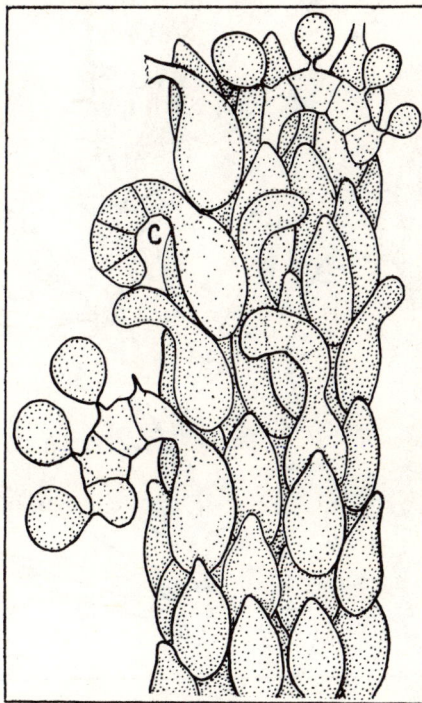

C. praelonga (Wint.) Arth.; teliospores.

36.   UNDESCRIBED (to be described by Buriticá and Hennen).

Spermogonia subepidermal, type 4?. Aecia and uredinia lacking. Telia subepidermal, deep-seated, becoming erumpent as brown, compact columns of adherent
spores; spores catenulate, with intercalary cells and an apparently common matrix,
1-celled, germinating without dormancy, germ pore not seen, basidium external.

TYPE: Cronartium jacksonii P. Henn.

One other species, on Xylopia aethiopica in Africa, is known. The genus differs from Cionothrix and Baeodromus in having intercalary cells and from Dietelia
in lacking a peridium.

REFERENCE: Buriticá, P. A revision of rust genera (Uredinales) with reduced
life cycles. Ph.D. thesis. Purdue University. 114 pp. 1974.

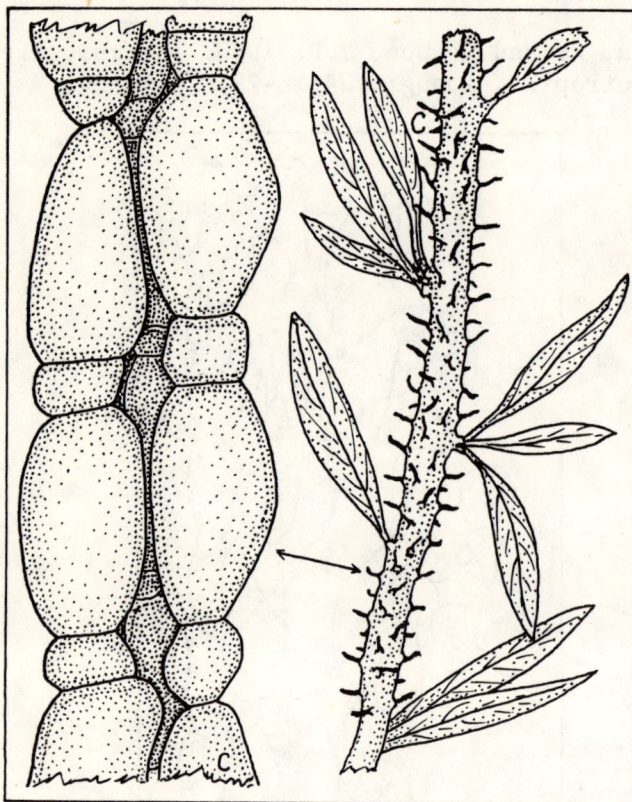

Undescribed; teliospores (left); habit on
host plant.

37. DIETELIA P. Hennings, Hedwigia 36:215. 1897.

Spermogonia subepidermal, type 4. Aecia and uredinia not produced. Telia subepidermal in origin, erumpent, with peridium, Aecidium-like but compact; spores
1-celled, catenulate, with intercalary cells basally but often not seen, wall pale
in most species, smooth or finely verrucose, germ pore 1 where seen, mostly obscure, basidium external.

TYPE: Dietelia verruciformis (P. Henn.) P. Henn. (Cronartium verruciforme
P. Henn.).

Six species are known, five from the Americas and one from the Philippines on
dicotyledonous plants. The species appear like Aecidium but are compact and the
peridium tends to adhere to the spore column. Endophyllum holwayi H.S. Jack. has
been treated by Lindquist as the monotypic genus Jacksoniella (Jacksonia Lindq.,
1970 not R. Brown 1811) but it has the characters of Dietelia. Endophylloides
Whet. & Olive also is treated as a synonym of Dietelia, following Buriticá and
Hennen.

REFERENCES: Buriticá, P. and Hennen, J.F. 1980. Pucciniosireae (Uredinales -
Pucciniaceae). Flora Neotropica, Monogr. 24:14-18. Lindquist, J.C. 1970. Notas
uredinologicas. Rev. Fac. Agron. La Plata. 46:199-205.

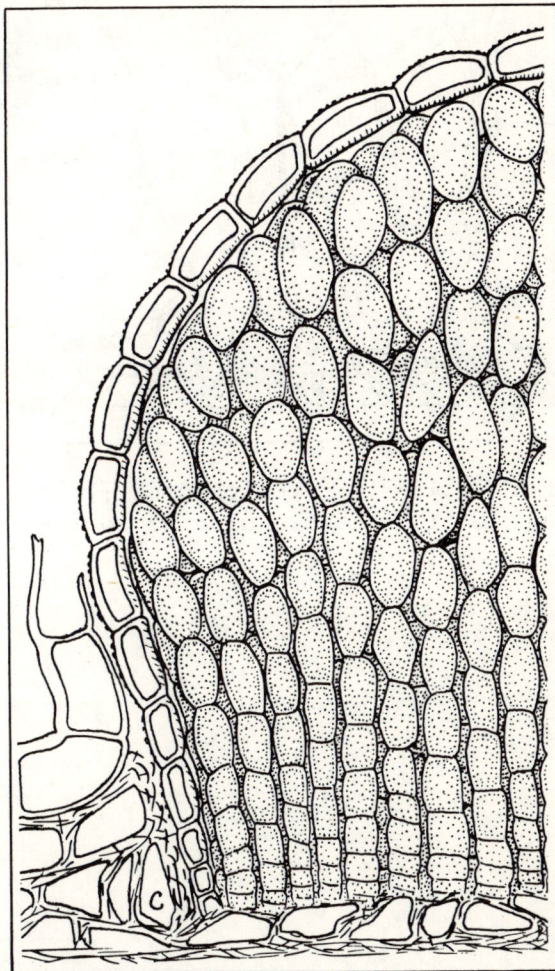

D. verruciformis (P. Henn.) P. Henn.
telium.

38. POLIOMA Arthur, J. Mycol. 13:29. 1907.

Spermogonia subepidermal, type 4. Aecia subepidermal in origin, erumpent, without peridium, caomatoid; spores catenulate, verrucose. Uredinia subepidermal in origin, erumpent, without peridium; spores borne singly, echinulate, pore 1, basal where known. Telia subepidermal in origin, erumpent, spores sessile and usually grouped on basal cells, 2-celled by horizontal septum, wall pale, germ pore 1 per cell, germination occurs without dormancy, basidium external.

TYPE: Polioma nivea (Holw.) Arth. (Puccinia nivea Holw.).

Four species have been assigned to this genus. Three are microcyclic on species of Salvia (Labiatae) and one macrocyclic species (P. reniformis León-Gall. & Cumm.). on Geranium. They occur in North and South America. Polioma differs from Puccinia because of its sessile teliospores.

REFERENCE: Baxter, J.W. and Cummins, G.B. 1951. Polioma Arth., a valid genus of the Uredinales. Bull. Torrey Bot. Club. 78:51-55.

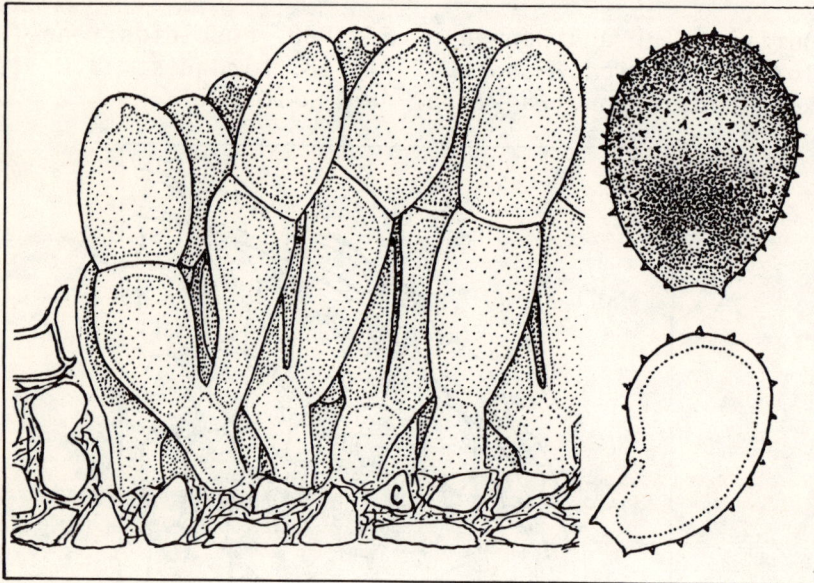

P. nivea (Holw.) Arth.; teliospores (left). P. reniformis Leoń-Gall. & Cumm.; urediniospores (right).

39. PUCCINIOSIRA Lagerheim, Ber. Dtsch. Bot. Ges. 9:344. 1891.

Spermogonia subepidermal, type 4 or lacking. Aecia and uredinia not produced. Telia subepidermal in origin, erumpent, with peridium, mostly Aecidium-like
but filiform in some species; spores 2-celled by horizontal septum, catenulate,
with intercalary cells, loosely or firmly united laterally and longitudinally,
wall pale or obviously pigmented, smooth or verrucose, the cells separating easily,
pores obscure, basidium external.

TYPE: Pucciniosira triumfettae Lagh.

The genus is composed of 17 species according to Buriticá, who treats Gambleola as a synonym. This seems radical at first thought, because of the long
dark telial columns of Gambleola, but the morphology is similar. Species of both
Pucciniosira sensu stricto and Gambleola occur on Berberis-Mahonia. Others parasitize a variety of dicotyledonous hosts, including Compositae.

REFERENCES: Buriticá, P. 1974. A revision of rust genera (Uredinales) with
reduced life cycles. Ph.D.thesis, Purdue University. 114 pp. Heim, R. 1953. Le
genre Pucciniosira Lagerh. Urediniana 3:15-20. Buriticá P. and Hennen, J.F. 1980.
Pucciniosireae (Uredinales - Pucciniaceae). Flora Neotropica, Monogr. 24:28-33.

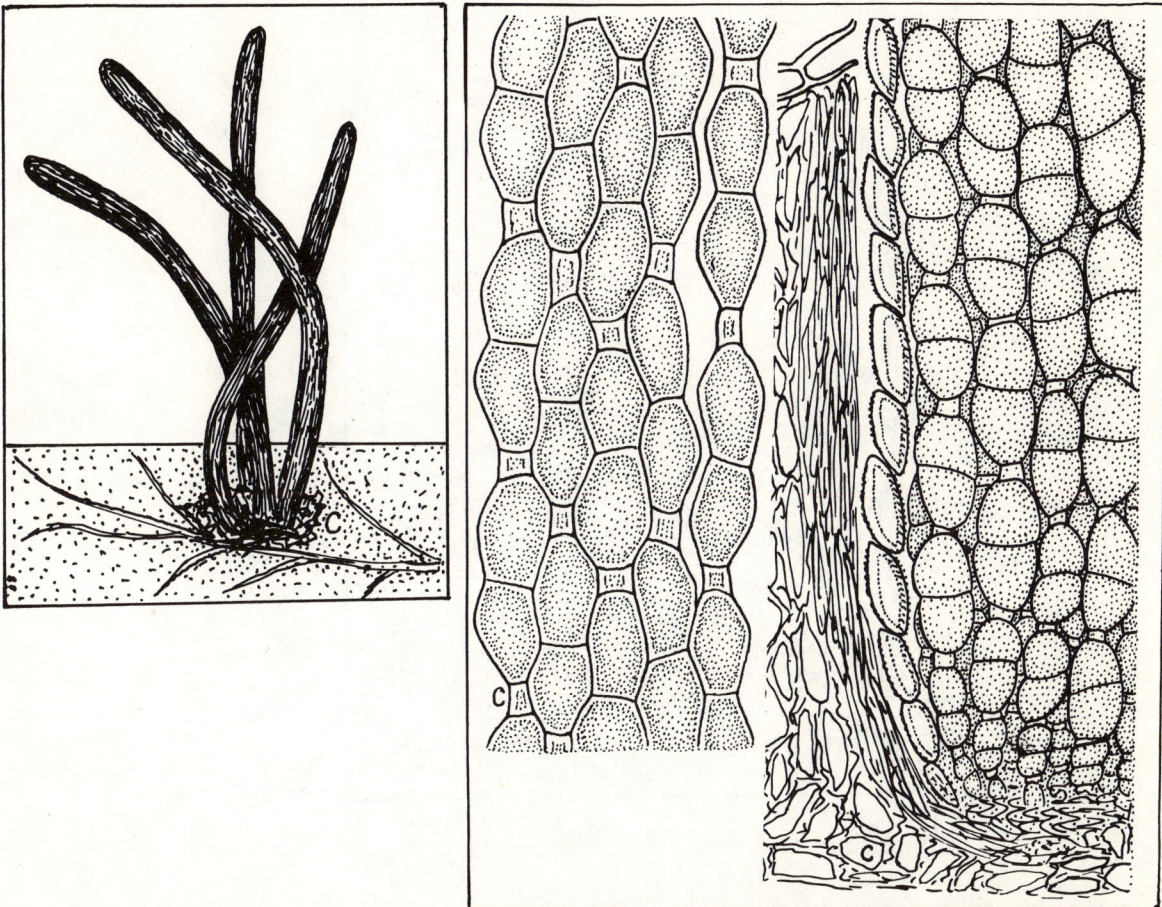

P. pallidula (Speg.) Lagh.; telia (right).
P. (Gambleola) cornuta (Massee); telial
habit and teliospores (left).

40. DIDYMOPSORA Dietel, Hedwigia 38:254. 1899.

Spermogonia subepidermal, type 4. Aecia and uredinia not produced. Telia subepidermal in origin, erumpent, dome-like or short columnar, somewhat like *Aecidium* in appearance but without peridium; spores 2-celled by horizontal septum, catenulate in loosely adherent chains, intercalary cells present but often seen only basally, wall pale, germ pores obscure but probably 1 per cell, germination occurs without dormancy, basidium external.

LECTOTYPE: *Didymopsora solani-argentei* (P. Henn.) Diet. (*Aecidium solani-argentei* P. Henn.).

Didymopsora and *Pucciniosira* differ only in the presence of a peridium in the latter. There are six species occurring in warm regions of Africa and the Americas on Compositae, Melastomataceae, Solanaceae, and Tiliaceae.

REFERENCES: Butiticá, P. and Hennen, J.F. 1980. Pucciniosireae (Uredinales - Pucciniaceae). Flora Neotropica, Monogr. 24:33-38. Cunningham, J.L. 1968. Ontogeny of teliospores of *Cronartium paraguayense* and relationship to *Didymopsora*. Mycologia 60:769-775.

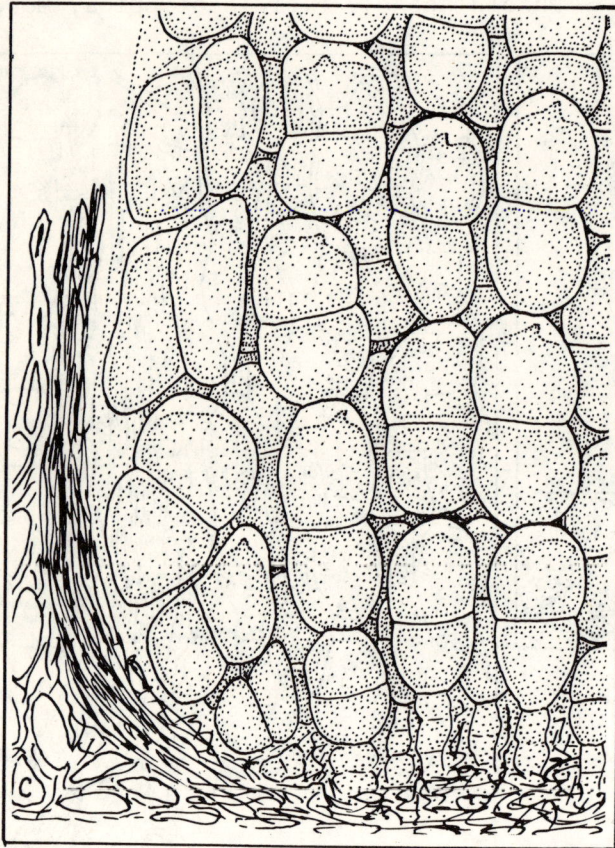

D. africana Cumm.; telium.

41. CHARDONIELLA Kern, Mycologia 31:375. 1939.

Spermogonia subepidermal, type 4. Aecia and uredinia not produced. Telia subepidermal in origin, becoming erumpent as hair-like columns of strongly adherent spores, the spores in no particular order in the column, 1-celled on long pedicel-like intercalary cells, germ pore 1 near apex, wall pale, germination occurs without dormancy, basidium external.

TYPE: Chardoniella gynoxidis Kern.

The telia appear as those of Cronartium but the spores are Uromyces-like. The columns are hard when dry. Buriticá and Hennen recognize four species, all on Compositae and all in South America. Further, they consider that the spores originate in chains with intercalary cells which elongate to simulate pedicels. Trichopsora is similar except for having an internal basidium.

REFERENCES: Buriticá, P. and Hennen, J.F. 1980. Puciniosireae (Uredinales - Pucciniaceae). Flora Neotropica, Monogr. 24:39-42. Buriticá, P. 1974. A revision of rust genera (Uredinales) with reduced life cycles. Ph.D. thesis. Purdue University. 114. pp.

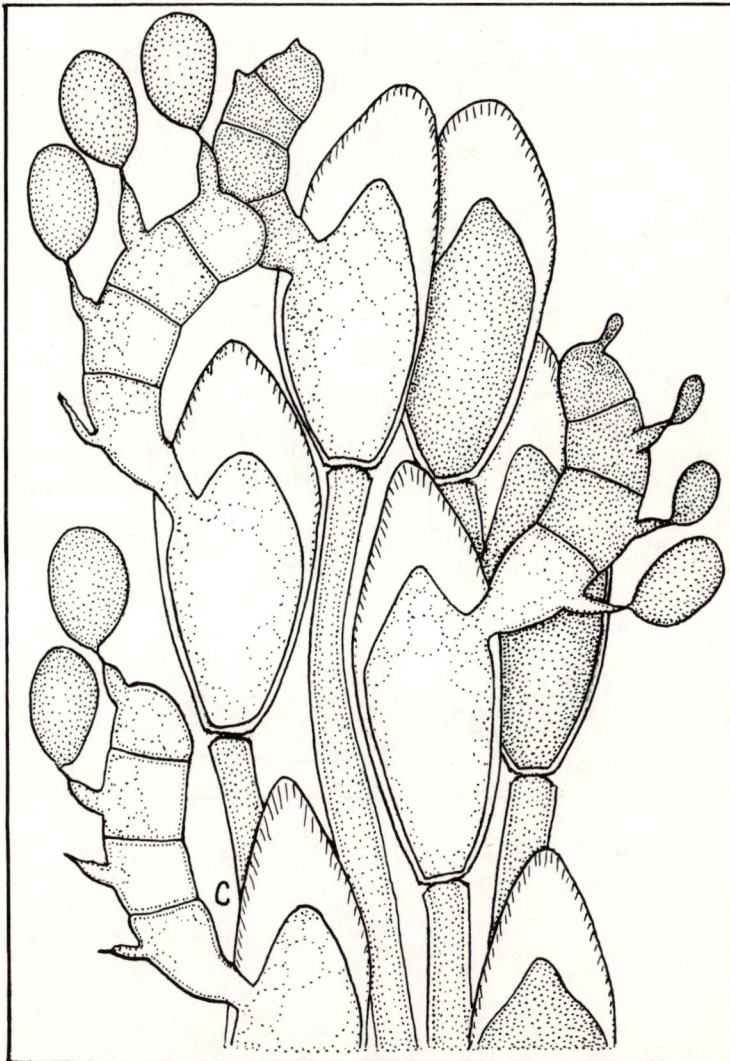

C. gynoxidis Kern; teliospores.

42. CHRYSELLA H. Sydow, Ann. Mycol. 24:292. 1926.

Spermogonia subepidermal, type 4. Aecia and uredinia unknown, probably not produced. Basidiosori subepidermal in origin, erumpent, waxy when wet, hard when dry; basidia borne singly on pedicels, becoming 4-celled by horizontal septa, each cell producing a long sterigma and a large basidiospore, germination occurs without dormancy.

TYPE: Chrysella mikaniae Syd.

This is the only species. It occurs on Mikania hirsutissima DC. in Costa Rica. Buriticá treats this species as belonging in Chrysocyclus.

REFERENCE: Buriticá, P. 1974. A revision of rust genera (Uredinales) with reduced life cycles. Ph.D. thesis. Purdue University. 114 pp.

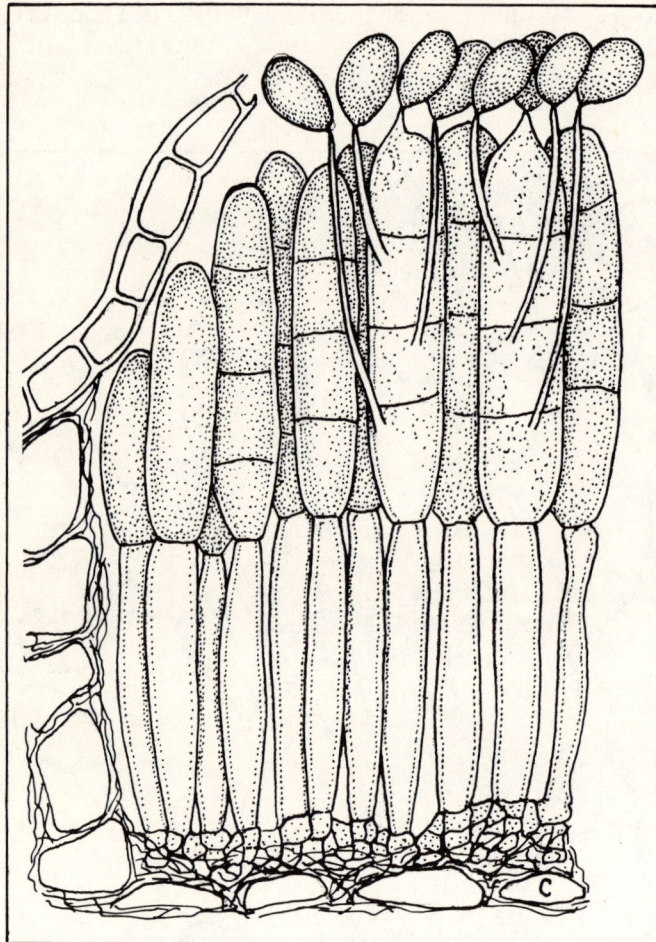

C. mikaniae Syd.; basidia and basidiospores.

43. ACHROTELIUM H. Sydow in Sydow and Petrak, Ann. Mycol. 26:425. 1928.

Spermogonia type 7. Aecia subepidermal in origin, erumpent, uredinoid, spores borne singly, echinulate. Uredinia subepidermal in origin, erumpent; spores borne singly, echinulate, pores obscure. Basidiosori subepidermal in origin, erumpent; basidia produced in groups on basal cells, pedicellate, wall pale, spores 1-celled at first becoming 4-celled by division of the protoplast, each cell producing 1 basidiospore, germination occurs without dormancy.

TYPE: Achrotelium ichnocarpi Syd.

There are five species on dicotyledonous hosts in warm regions.

REFERENCE: Ono, Y. 1978. Taxonomic revisions of the tribe Oliveae and morphologically related genera (Uredinales). Ph.D. thesis, Purdue University. 254 pp.

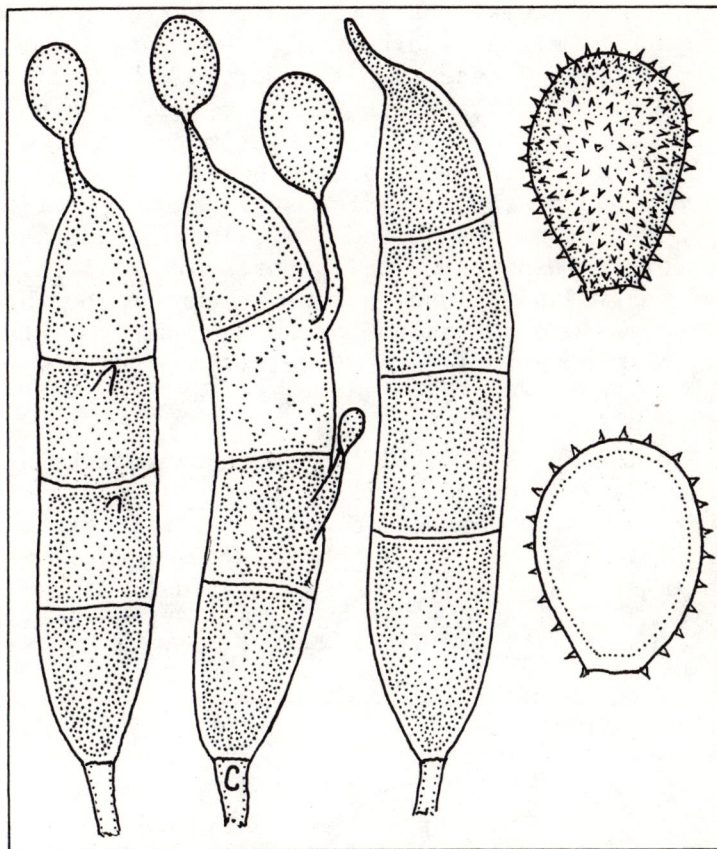

A. ichnocarpi Syd.; basidia and urediniospores.

44.   HEMILEIA Berkeley & Broome, Gard. Chron. 1869:1157. 1869.

Spermogonia and aecia unknown. Uredinia typically suprastomatal but in some
species erumpent through the ruptured epidermis; spores borne singly on short pedicels, typically strongly asymmetrical with 1 flat or concave side, usually
smooth, and a dorsal, convex, echinulate or aculeate side. Telia as the uredinia
of the species; spores borne singly on short pedicels, 1-celled, usually broader
than high, often angular, wall uniformly thin, pale, germ pore not visible, perhaps not differentiated, germination occurs without dormancy; basidium external.

TYPE: Hemileia vastatrix Berk. & Br.

About 50 species have been described, most from Africa and Asia. The "hump-
backed" urediniospores are so characteristic that numerous species have been described without telia being known. Despite many studies and inoculations, no
aecial state is known (see Thirumalachar). Because the teliospores germinate
without dormancy, the aecial state is expected to be systemic or otherwise persistent. But, in H. vastatrix, Rajendren has reported a "...unique nuclear behavior, with karyogamy and meiosis located in the so-called urediospores...."
There is question whether Hemileia is a legitimate name. Only urediniospores
are described. The illustration is of clustered urediniospores, as seen from the
upper side of the sorus, and a single urediniospore, and a "... group of threads
with young immature spores highly magnified ...." The threads and spores bear no
resemblance to either urediniospores or teliospores of Hemileia. But, because
Hemileia is a name of long-standing and because it cannot be stated categorically
that the threads and spores are not teliospores, the name is retained.
H. vastatrix is an important disease of cultivated coffee. Species of the
genus parasitize both monocot and dicot hosts.

REFERENCES: Gopalkrishnan, K.S. 1951. Notes on the morphology of the genus
Hemileia. Mycologia 43:271-283. Harr, J. 1979. Hemileia vastatrix Berk. & Br.
Publ. Div. Agron. Dept. Investig. Sandos, S.A. 27 pp. Rajendren, R.B. 1967. A new
type of nuclear life cycle in Hemileia vastatrix. Mycologia 59:279-285. Rayner, R.
R.W. 1972. Micologia, historia y biologia de la roya del cafeto. Publ. Misc. No.
94, Inst. Interamer. Cienc. Agric. O.E.A., Turialba, Costa Rica, 68. pp. Thirumalachar, M.J. Some noteworthy rusts - II. Mycologia 39:231-248. 1947.

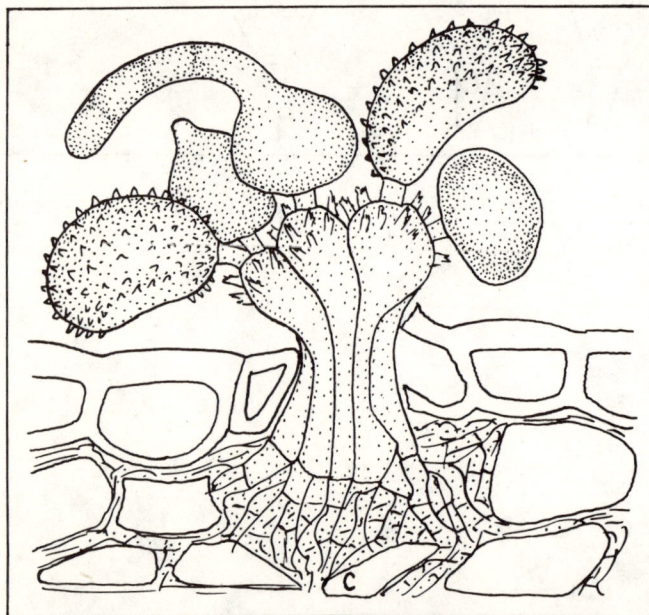

H. vastatrix Berk. & Br.; sorus with
urediniospores and teliospores.

45. GERWASIA Raciborski, Bull. Acad. Sci. Cracovie Cl. Sci. Math. Nat. 1909:207.

Spermogonia within the hypertrophied epidermis, type 6. Aecia also intraepidermal, uredinoid; spores borne singly on pedicels, echinulate. Uredinia intraepidermal or subepidermal in origin, becoming erumpent; spores borne singly on pedicels, echinulate, pores obscure, perhaps equatorial. Telia as the uredinia or strictly suprastomatal; spores borne singly on pedicels, 1-celled, wall pale, germ pore 1, germination occurs without dormancy, basidium external.

TYPE: Gerwasia rubi Racib.

Raciborski's type has not been seen but there is no doubt that the genus is the same as Mainsia H.S. Jack., and takes precedence over it. It is probable that Mainsia rubi (Diet. & Holw.) H.S. Jack. is the same fungus as G. rubi because both have suprastomatal telia. Most of the 16 or so species of Mainsia have not been transferred. Assigning species with suprastomatal, intraepidermal, or subepidermal sori may seem anomalous but similar treatment is accorded Hemileia and Prospodium. The species occur on Rosa and more especially on Rubus. All species are autoecious and most occur in the Asian and American tropics.

REFERENCE: Jackson, H.S. 1931. The rusts of South America based on the Holway Collections - III. Mycologia 23:96-116.

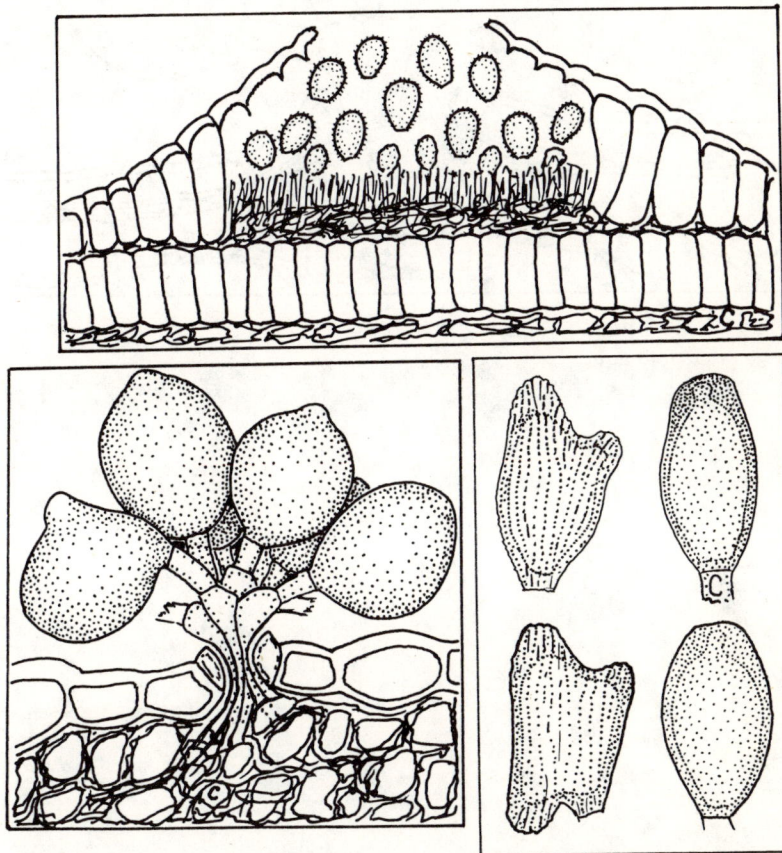

G. rubi Racib.; uredinium (above), telium (left);
G. standleyi (Cumm.) Cumm. & Y. Hirat. comb. nov.;
Mainsia standleyi Cumm.); urediniospores and teliospores (right).

46. BLASTOSPORA Dietel, Ann. Mycol. 6:222. 1908.

Spermogonia and aecia unknown. Uredinia subepidermal in origin but supra-stomatal whem mature, the spores produced from sporogenous cells that emerge through stomata; spores borne singly, echinulate, pores obscure. Telia suprastom-atal as the uredinia; spores 1-celled, borne singly on pedicels, wall thin and pale, without a clearly differentiated germ pore, germination occurs without dor-mancy by growth of the apical region of the spore, basidium external.

TYPE: Blastospora smilacis Diet.

REFERENCES: Kaneko, S. and Hiratsuka, N. 1981. Blastospora betulae (Uredin-ales) from Japan. Mycologia 73:577-580. Mains, E.B. 1938. The genus Blastospora. Amer. J. Bot. 25:677-679.

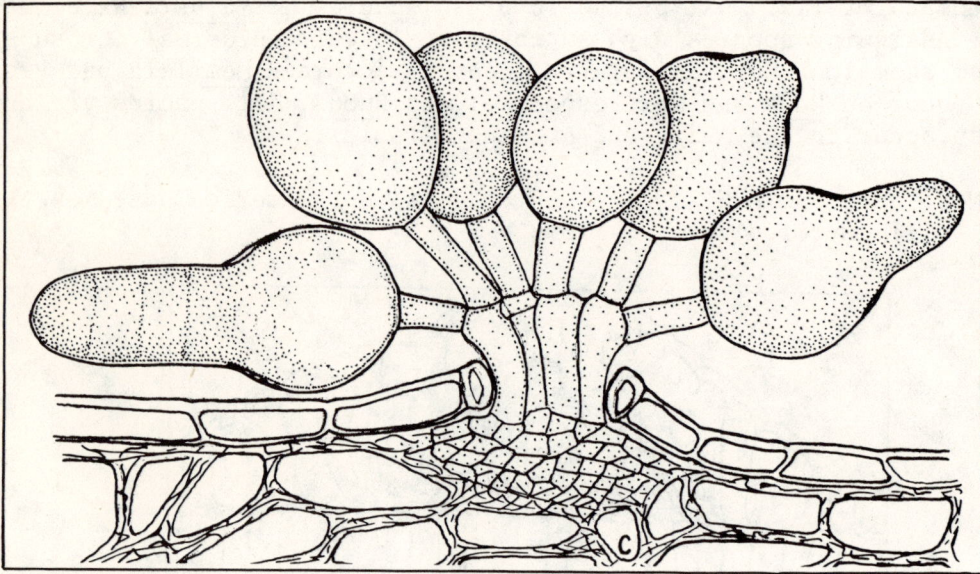

B. itoana Toga. & Onuma; teliospores.

47.   ZAGHOUANIA Patouillard, Bull. Soc. Mycol. France 17:187. 1901.

Spermogonia subepidermal, type 4. Aecia subepidermal in origin, erumpent, with or without peridium but aecia in cup-like cavaties; spores catenulate in both types. Uredinia subepidermal in origin, erumpent; spores borne singly, echinulate. Telia subepidermal in origin, either rupturing the epidermis or developing suprastomatally; spores 1-celled, borne singly on pedicels, wall thick, pale, and verrucose, germination occurs without dormancy, basidium external, short, thick, and rather thick-walled, basidiospores sessile, not ejected.

TYPE: Zaghouania phillyreae Pat.

Two species are recognized. The type has erumpent telia; Z. oleae has suprastomatal telia. Z. oleae often is treated as Cystopsora oleae Butl. but maintenance of two genera does not seem justified. Z. phillyreae has 4-celled basidia but Z. oleae only two basidiospores.

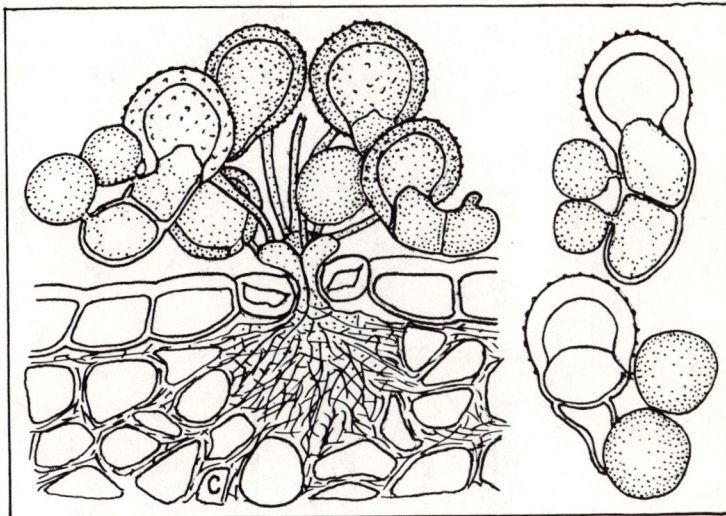

Z. oleae (Butl.) Cumm.; teliospores.

48.   BOTRYORHIZA Whetzel & Olive, Amer. J. Bot. 4:47. 1917.

Spermogonia, aecia, and uredinia unknown. Telia subepidermal in origin, becoming erumpent, pulverulent; spores 1-celled, borne singly on pedicels, wall pale and thin, without a differentiated germ pore, germination occurs without dormancy by continued growth of the apex of the spore, basidium external.

TYPE: Botryorhiza hippocrateae Whet. & Olive.

Botryorhiza is a monotypic genus that occur is the West Indies and Brazil. Ono treats the genus as a synonym of Maravalia but Botryorhiza has priority. Moreover, Botryorhiza is poorly understood and is best retained as monotypic.

REFERENCE: Ono, Y. 1978. Taxonomic revisions of the tribe Oliveae and morphologically related genera (Uredinales). Ph.D. thesis. Purdue University. 254. pp.

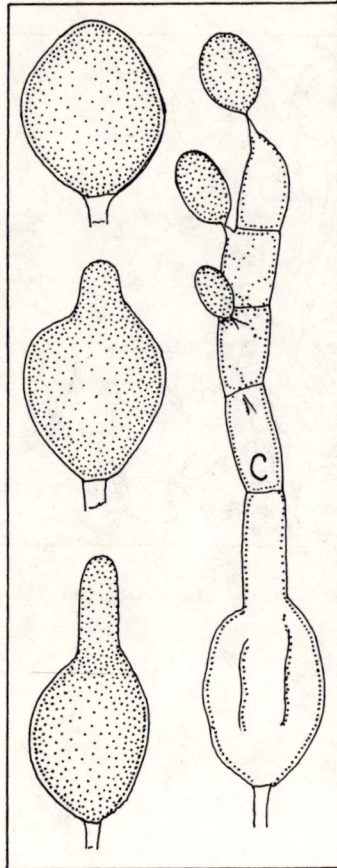

B. hippocrateae Whet.
& Olive; teliospores.

49. MARAVALIA Arthur, Bot. Gaz. 73:60. 1922.

Spermogonia subcuticular, type 5 or 7 (but unknown in the type). Aecia sub-epidermal or subcuticular in origin, erumpent, with or without paraphyses, uredinoid; spores borne singly on pedicels, mostly echinulate, pores various, often unizonate. Uredinia subepidermal in origin, erumpent, similar to the aecia; spores as the aeciospores. Telia subepidermal in origin, erumpent, spores 1-celled, borne singly on pedicels or occasional spores sessile, with or without sporogenous basal cells, wall thin, pigmented or pale, germ pore differentiated or not, germination apical and without dormancy, basidium external.

TYPE: Maravalia pallida Arth. & Thaxt.

Maravalia, as often treated, is heterogeneous because species with type 4 spermogonia are included. When such species are removed, as we do here, Maravalia, Scopella, and Angusia have most features in common; all are tropical, all have pedicellate, non-resting teliospores, all have uredinoid aecia or probably do, and, insofar as known, all have type 5 or 7 spermogonia. So we reduce Scopella and Angusia to synonymy. Ono treated Angusia as a synonym of Scopella, but perhaps because he considered that Maravalia had type 4 spermogonia, he did not merge them with Maravalia.

REFERENCES: Cummins, G.B. 1950. The genus Scopella of the Uredinales. Bull. Torry Bot. Club 77:204-213. Laundon, G.F. 1964. Angusia (Uredinales). Trans. Brit. Mycol. Soc. 47:327-329. Mains, E.B. 1939. Scopella gen. nov. of the Pucciniaceae. Ann. Mycol. 37:57-60. Mains, E.B. 1939. Studies in the Uredinales, the genus Maravalia. Bull. Torrey Bot. Club 66:173-179. Ono, Y. 1978. Taxonomic revisions of the tribe Oliveae and morphologically related genera (Uredinales). Ph.D. thesis. Purdue University. 254 pp. Thirumalachar, M.J. 1950. Some noteworthy rusts. III. Mycologia 42:224-232.

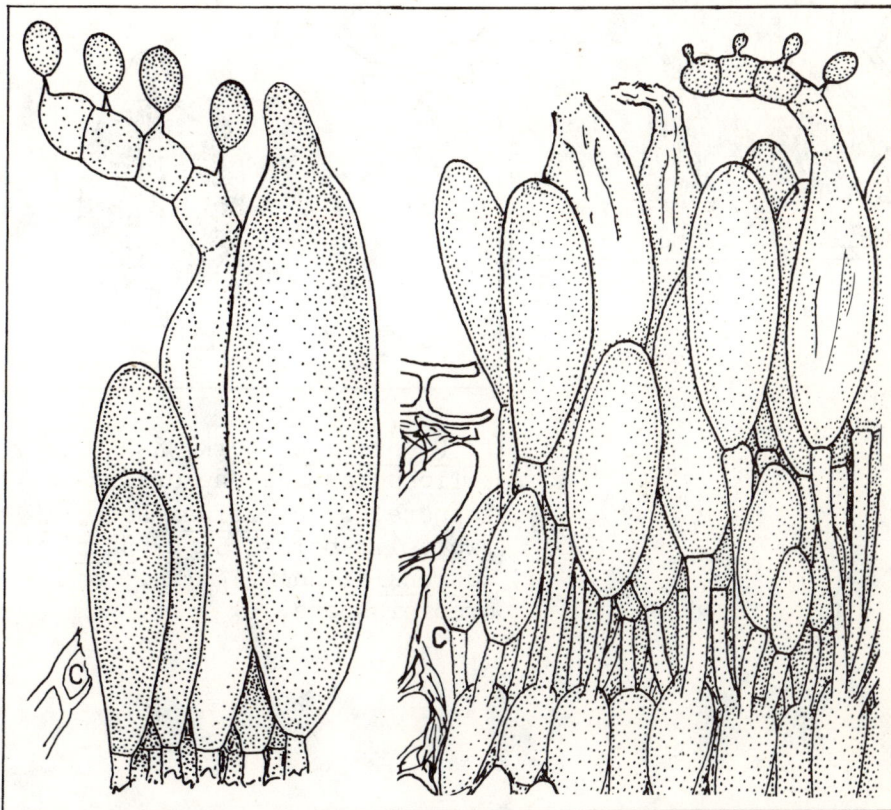

M. pallida Arth. & Thaxt. (left). M. kevorkianii (Cumm.) Cumm. & Y. Hirat. comb. nov. (Scopella kevorkianii Cumm.); teliospores (right).

50. PILEOLARIA Castagne, Obs. Pl. Acotyl. Fam. Ured. 1:22. 1842.

Spermogonia subcuticular, type 7. Aecia subepidermal in origin, erumpent, uredinoid; spores borne singly on pedicels, verrucose or echinulate, often in spiral or longitudinal lines, pores zonate. Uredinia subepidermal in origin, erumpent; spores as the aeciospores. telia subepidermal in origin, erumpent; spores 1-celled, borne singly on pedicels, wall thick and pigmented, variously sculptured, germ pore 1, germination mostly after dormancy, basidium external.

TYPE: Pileolaria terebinthi Cast.

There are some 20 to 24 species, most of which are macrocyclic. They parasitize species of Rhus and Pistacia of the Anacardiaceae. Teliospores are typically somewhat discoid and verrucose or reticulately verrucose, and the urediniospores are ridged or beaded in longitudinal or spiral patterns. A few species on Leguminosae have been referred to Pileolaria but we treat these as species of Atelocauda.

REFERENCE: Katsuya, K., Kakashima, M. and Sato, S. 1980. Spore surface structure of three Pileolaria species in Japan. Rept. Tottori Mycol. Inst. Japan 18:163-167. Excellent SEM micrographs.

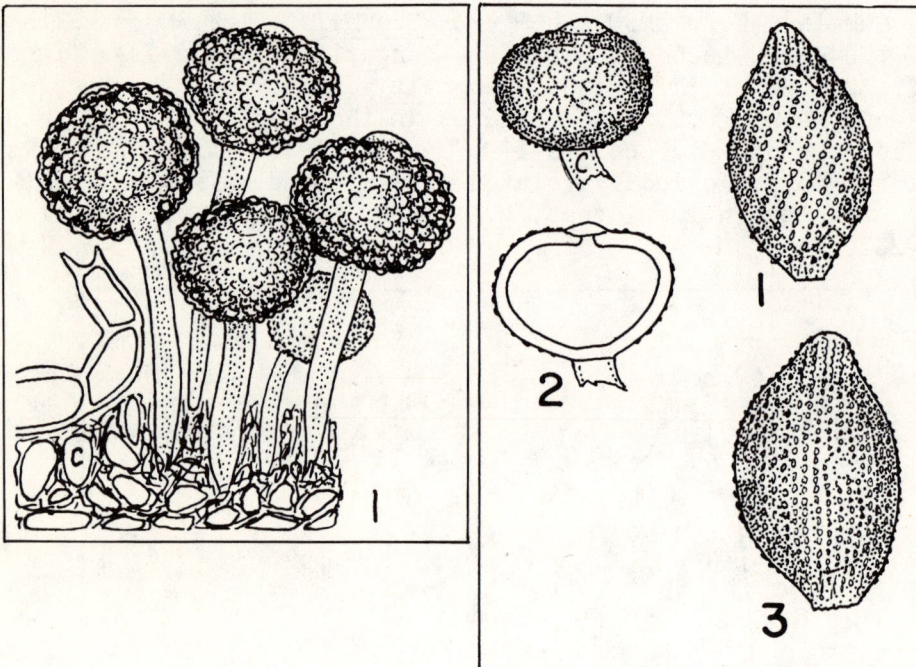

1. P. brevipes Berk. & Rav.; teliospores and one urediniospore. 2. P. effusa Peck; two teliospores. 3. P. patzcuarensis (Holw.) Arth.; one urediniospore.

51. ATELOCAUDA Arthur & Cummins, Ann. Mycol. 31:41. 1933.

Spermogonia subepidermal or subcuticular, type 5 or 7. Aecia subepidermal in origin, erumpent, uredinoid; Spores borne singly on pedicels, similar to the urediniospores. Uredinia subepidermal in origin, erumpent; spores borne singly on pedicels, pigmented, most and perhaps all reticulate, pores zonate. Telia subepidermal in origin, erumpent; spores 1-celled, borne singly on pedicels, wall pale or pigmented, germ pore 1, pedicels simple, basidium external.

TYPE: Atelocauda incrustans Arth. & Cumm.

The type, on Lonchocarpus in Panama, is microcyclic. Others are macrocyclic and occur on Acacia in Australia, New Zealand, and Hawaii. They are A. bicincta (McAlp.) comb. nov. (Uromyces bicinctus McAlp.), A. koae (Arth.) comb. nov. (U. koae Arth.), A. digitata (Wint.) comb. nov. (U. digitatus Wint.). Atelocauda is readily separable from Uromyces because of the spermogonia. It is less clearly separable from Pileolaria but it is doubtful that the relationship is with Pileolaria. Atelocauda more likely is related to Uromycladium.

REFERENCES: Gardner, D.E. 1981. Nuclear behavior and clarification of the spore stages of Uromyces koae. Can. J. Bot. 59:939-946. Thirumalachar, M.J. and Kern, F.D. 1955. The rust genera Allotelium, Atelocauda, Coniostelium and Monosporidium. Bull. Torrey Bot. Club 82:102-107.

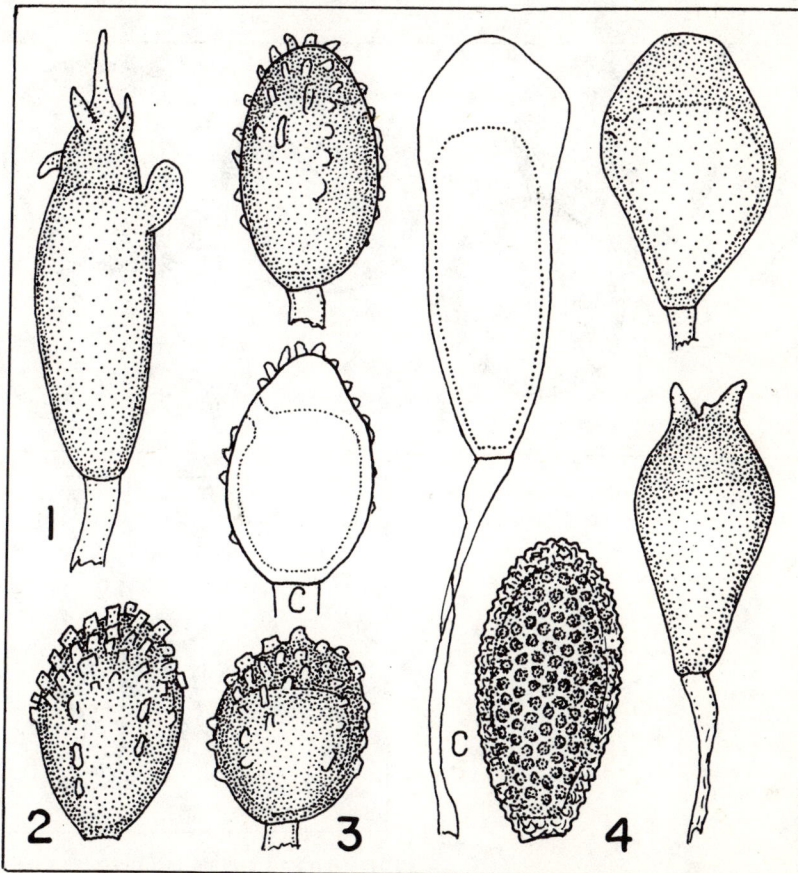

1. A. digitata; one teliospore. 2. A. incrustans; one teliospore. 3. A. koae; three teliospores. 4. A. bicincta; teliospores and one urediniospore.

52.   UROMYCLADIUM McAlpine, Ann. Mycol. 3:321. 1905.

Spermogonia subepidermal, type 5 or perhaps sometimes type 7. Aecia sub-epidermal in origin, erumpent, uredinoid; spores borne singly on pedicels, verrucose or reticulate, pores equatorial. Uredinia subepidermal in origin, erumpent; spores as the aeciospores. Telia subepidermal in origin, erumpent; spores 1-celled with 1 to 3 borne on a usually branched and septate pedicel which may bear hygroscopic spore-like cysts, germ pore 1, the spore usually pigmented, basidium external.

LECTOTYPE: Uromycladium simplex McAlp.

Seven species have been recognized, occurring on Acacia and Albizzia of the Leguminosae. Both macrocyclic and microcyclic species are known; all are auto-ecious. Some species cause galls and distortions of the host plant. Perhaps also related to Uromycladium is Atelocauda. Some species of Atelocauda produce distortions and have urediniospores with surface marking similar to Uromycladium. These species have type 5 spermogonia, as does Uromycladium but they have nonseptate pedicels and no sterile cells.

REFERENCES: Cunningham, G.H. 1931. The rust fungi of New Zealand. 261 pp. McIndoe, Dunedin, N. Z. McAlpine, D. 1905. A new genus of Uredineae -Uromycladium. Ann. Mycol. 3:303-323. McAlpine, D. The rusts of Australia. 349 pp. Govt. Printer, Melbourne.

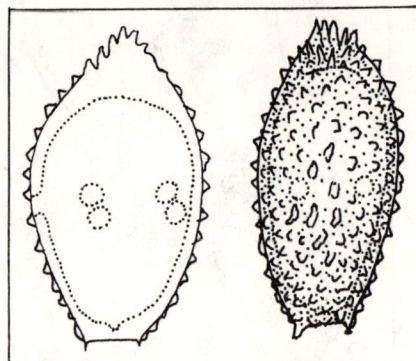

U. maritimum McAlp.; telio-spores and urediniospores.

53. LIPOCYSTIS Cummins, Bull. Torrey Bot. Club 64:39. 1937.

Spermogonia subcuticular, type 7. Aecia subcuticular in origin, erumpent, uredinoid; spores borne singly on pedicels, echinulate, pores equatorial. Uredinia subcuticular in origin, erumpent; spores as the aeciospores. Telia subcuticular in origin, erumpent; spores 1-celled, borne singly on short, broad pedicels, laterally free or tending to adhere in small groups, wall pigmented, germ pore 1, basidium external.

TYPE: Lipocystis caesalpiniae (Arth.) Cumm. (Uromyces caesalpiniae Arth., lectotype Stevens 393).

Only this species is known. It is autoecious and occurs in the West Indies on Mimosa. It is one of the few genera that have all spore forms subcuticular. There is little to indicate relationship except the host and the spermogonia, both of which suggest Ravenelia.

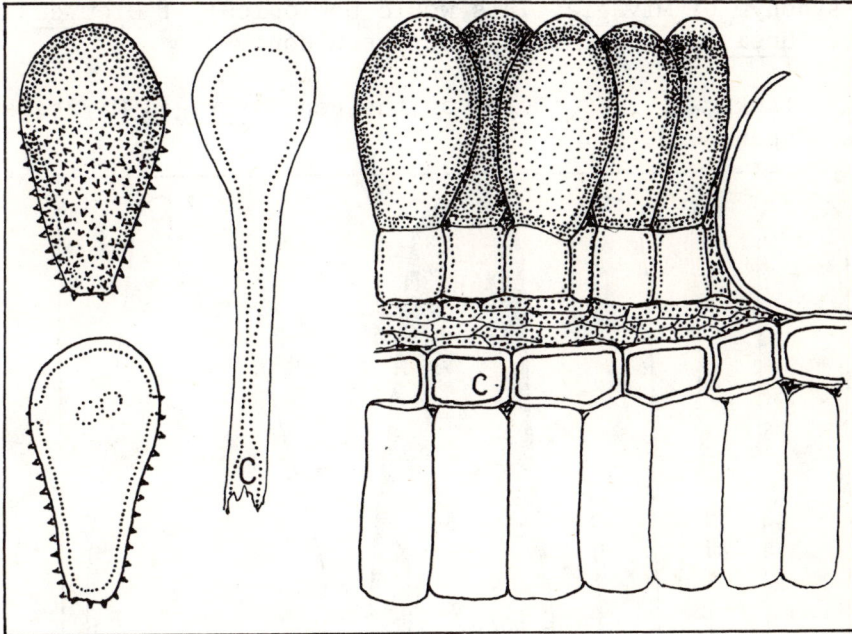

L. caesalpiniae (Arth.) Cumm.; urediniospores, one paraphysis, and teliospores.

54. CORBULOPSORA Cummins, Mycologia 32:364-365. 1940.

Spermogonia deep-seated, type 4. Aecia deep-seated, becoming erumpent, aecidioid, with peridium; aeciospores catenulate. Uredinia subepidermal in origin,
becoming erumpent, with a tubular peridium composed of long, laterally united,
palisade-like cells; spores borne singly on pedicels, echinulate, pores scattered
or obscure. Telia subepidermal in origin, erumpent, with a peridium as the uredinia; spores 1-celled, borne singly on pedicels, several of which arise from single basal sporogenous cells, germ pore 1, basidium external.

TYPE: Corbulopsora clemensiae Cumm.

Three species have been recognized, all on Compositae, the type and C. grav-
ida Cumm. on Olearia in New Guinea, and C. cumminsii Thir. on Lactuca in India.
Corbulopsora differs from Uromyces because of the palisade-like peridium, just as
Miyagia differs from Puccinia. In the Illustrated Genera of 1959, Corbulopsora
was treated as a synonym of Miyagia. This would be logical if Uromyces were treated as a synonym of Puccinia, but this has not been done.

REFERENCE: Hiratsuka, N. 1969. Notes on the genus Miyagia Miyabe ex Sydow.
Trans. Mycol. Soc. Japan 10:89-90.

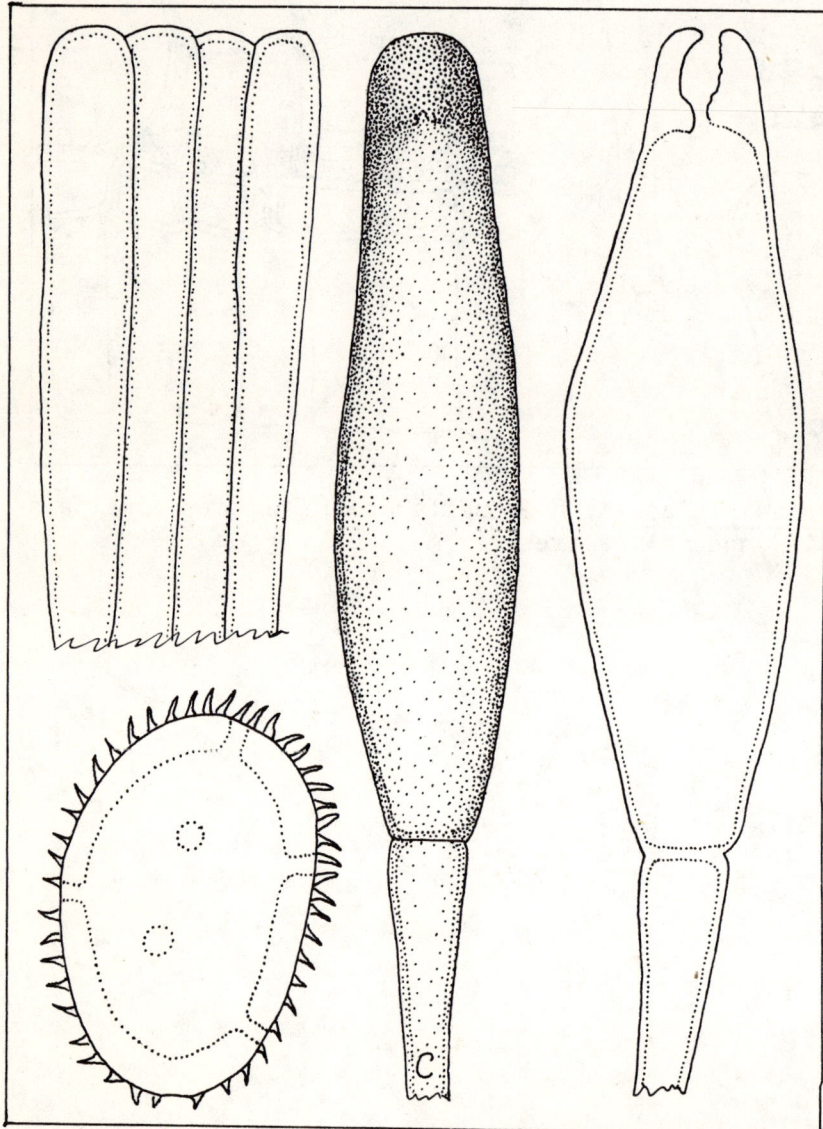

C. clemensiae Cumm.; peridial cells and spores.

55. UROMYCES (Link) Unger, Einfluss Bodens p. 216. 1836.

Spermogonia subepidermal, type 4. Aecia subepidermal in origin, erumpent, either aecidioid with peridium and catenulate spores or uredinoid with spores borne singly on pedicels. Uredinia subepidermal in origin, erumpent; spores borne singly on pedicels, usually echinulate, pores various, mostly obvious. Telia subepidermal in origin, erumpent or remaining covered by the epidermis; spores borne singly on pedicels which may or may not be grouped on basal sporogenous cells, wall mostly pigmented, pore 1, basidium external, spores 1-celled.

LECTOTYPE: Uromyces appendiculatus (Pers.) Unger.

Uromyces is the second largest genus of rust fungi. Its species parasitize monocots and dicots throughout the world. The Compositae, Gramineae, Liliaceae, Euphorbiaceae, and Leguminosae support more species than other families. All variations of life cycles occur. For whatever reason, there are many species of Uromyces on legumes but very few species of Puccinia. Uromyces differs from Puccinia only in having 1-celled teliospores but is maintained for convenience and historical reasons. To treat Uromyces as a synonym of Puccinia would entail many nomenclatural changes with no real benefit.

There are several economically important species: bean rust (U. appendiculatus), pea rust (U. pisi), alfalfa or lucern rust (U. striatus), clover rust (U. trifolii-repentis), beet rust (U. betae), chickpea rust (U. ciceris-arietinus), and carnation rust (U. dianthi).

REFERENCES: Azbukina, Z.M. 1974. Rust fungi of the Soviet Far East (in Russian). Akad. Nauk, Moscow. 527 pp. Cummins, G.B. 1971. The rust fungi of cereals, grasses and bamboos. Springer-Verlag. 570 pp. Cummins, G.B. 1978. Rust Fungi on legumes and composites in North America. Univ. Arizona Press. 424 pp. Gaeumann, E. Die Rostpilze Mitteleuropas. Buechler & Co., 1407 pp. Guyot, A.L. 1938, 1951, 1957. Les Urédinées. Genre Uromyces. Encycl. Mycol. Vol. VIII. 438 pp. Vol. XV. 331 pp., Vol. XXIX. 649 pp. Paul Lechevalier. Paris. Hiratsuka, N. 1973. Revision of taxonomy of the genus Uromyces in the Japanese Archipelago. Rept. Tottori Mycol. Inst. Japan. 10:1-98. Lindquist, J.C. 1982. Royas de la Republica Argentina y zonas limitrfes. Coleccion Cient. XX:120-212.

Illustrations next page.

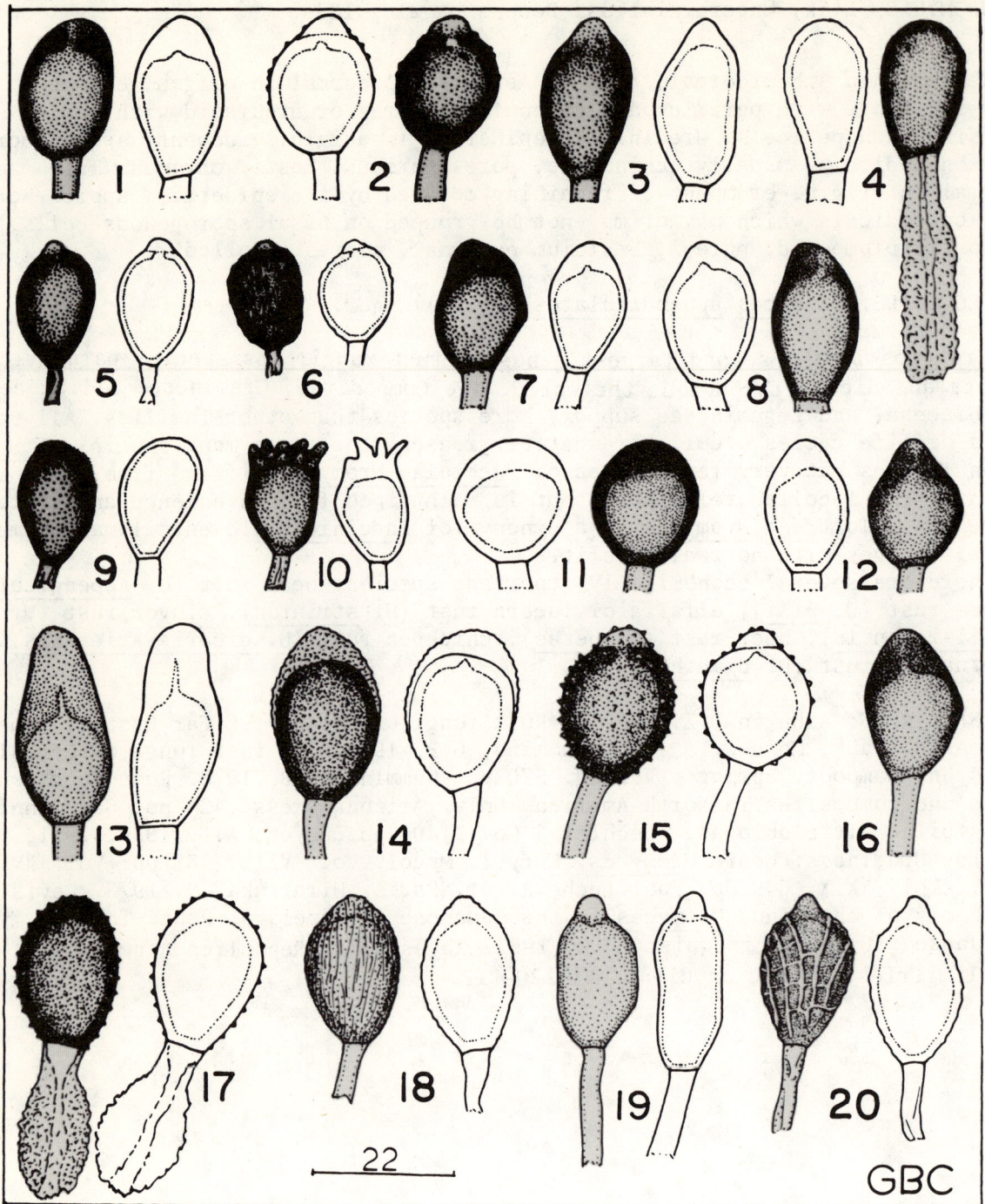

Uromyces species; teliospores. 1, viciae-fabae Schroet. 2, appendiculatus (Pers.)Unger. 3, vignae Barcl. 4, amurensis Kom. 5, glycyrrhizae Magn. 6, striatus Schroet. 7, eragrostidis Tracy. 8. aleuropodis-repentis Nattr. 9. dactylidis Otth. 10, halstedii DeT. 11, linearis Berk. & Br. 12, seditiosus Kern. 13, gemmatus Berk. & Curt. 14, corollocarpi Dale. 15, cucullatus Syd. 16, commelinae Cke. 17, sonorensis Hennen & Cumm. 18, socius Arth. & Cumm. 19, bidenticola Arth. 20, erythronii Pass.

56. CHRYSOPSORA Lagerheim, Ber. Dtsch. Bot. Ges. 9:345. 1891 issued 1892.

Spermogonia subepidermal, type 4. Aecia and uredinia not produced. Basidiosori subepidermal in origin, erumpent, waxy when fresh or moist; basidia borne in pairs on pedicels, separated by a horizontal septum, each cell of the pair becoming 4-celled by division of the protoplast, wall pale, germination occurs without dormancy.

TYPE: Chrysopsora gynoxidis Lagh.

The type and only species occurs on Gynoxis of the Compositae in South America. There are no good clues to the relationship of this genus but a point of interest is that it proves that an "internal basidium" alone cannot characterize the Coleosporiaceae.

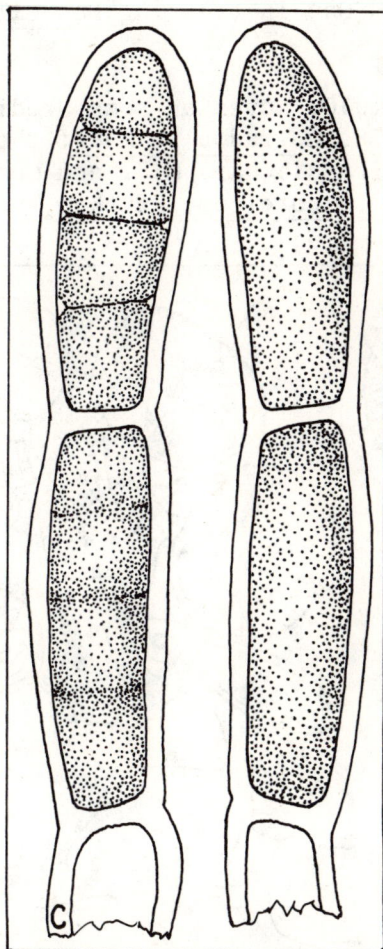

C. gynoxidis Lagh.;
basidia.

57. DESMELLA H. Sydow & P. Sydow, Ann. Mycol. 16:241. 1918 (issued 1919).

Spermogonia and aecia unknown. Uredinia suprastomatal as the telia; spores borne singly on pedicels, echinulate. Telia substomatal in origin, emerging from stomates by means of a few sporogenous cells, each of which may produce several spores, the mature sorus thus suprastomatal; spores borne singly on pedicels, 2-celled by horizontal, oblique, or vertical septum, wall pale or pigmented, germ pore 1 in each cell, germinating without dormancy; basidium external.

TYPE: Desmella aneimiae Syd.

There are three or four species of uncertain distinction, all on ferns. The relationship of the genus will remain obscure at least until the aecial state is known. Edythea has been treated as synonymous with Desmella (see Thirumalachar and Cummins) but we follow Hennen and Ono in treating Desmella as distinct (see discussion under Edythea).

REFERENCES: Cummins, G.B. 1940. Notes on some Uredinales. Ann. Mycol. 38:335-338. Hennen, J.F. and Ono, Y. 1978. Cerradoa palmaea: the first rust fungus on Palmae. Mycologia 70:569-576. Thirumalachar, M.J. and Cummins, G.B. 1948. Status of the rust genera Allopuccinia, Leucotelium, Edythea, and Ypsilospora. Mycologia 40:417-422.

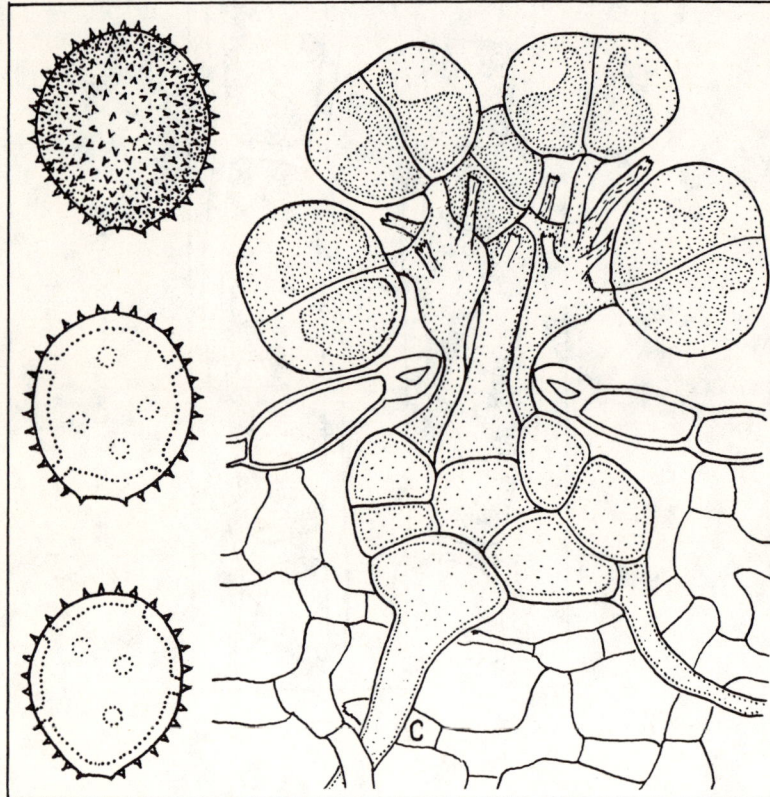

D. aneimiae Syd.; urediniospores and teliospores.

58. EDYTHEA H.S. Jackson, Mycologia 23:97. 1931.

Spermogonia and aecia unknown. Uredinia suprastomatal like the telia; spores borne singly on pedicels, echinulate. Telia subepidermal in origin but emerging through stomata by means of elongate cells which produce sporogenous cells apically which, in turn, produce spores borne singly on pedicels, 2-celled, septum vertical or horizontal, germ pore 1 in each cell; basidium external.

TYPE: Edythea quitensis (Lagerh.) H.S. Jack. & Holw. (Uropyxis quitensis Lagerh.).

The type and two other species occur on Berberis in South America. Cerradoa palmaea Hennen & Ono, which parasitizes Attalea ceraensis Barb. Rodr. of the Palmae, differs in that the teliospores are deeply pigmented. To accept degrees of pigmentation of spore walls as generic characteristics is hardly tenable. Consequently, Cerradoa palmaea is here reduced to synonymy as Edythea palmaea (Hennen & Ono) comb. nov. Edythea in turn, differs from Desmella because in Desmella the teliospores arise on the apices of elongate sporogenous cells which emerge through the stomata. Whether this is a significant distinction is questionable. Of greater significance may be the type of spermogonia of these genera and none of the primary stages is known.

REFERENCES: Hennen, J.F. and Ono, Y. 1978. Cerradoa palmaea: the first rust fungus on Palmae. Mycologia 70:569-576. Thirumalachar, M.J. and Cummins, G.B. 1948. Status of Allopuccinia, Leucotelium, Edythea, and Ypsilospora. Mycologia 40: 417-422.

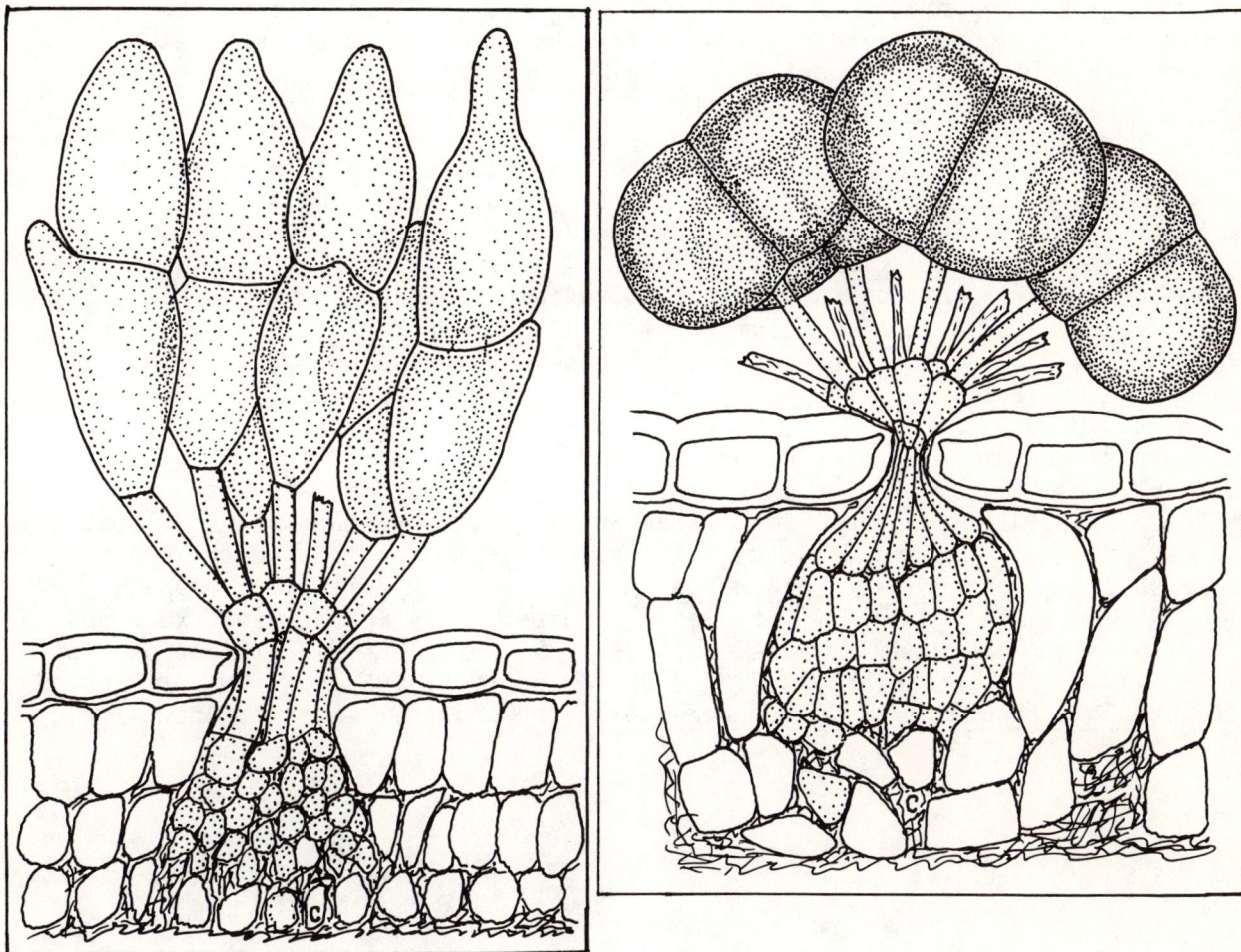

E. tenella H.S. Jack. & Holw.; teliospores (left). E. palmaea (Hennen & Ono) Cumm. & Y. Hirat.; teliospores (right).

59. GYMNOSPORANGIUM Hedwig f. in DeCandolle, Fl. France 2:216. 1805.

Spermogonia subepidermal, type 4. Aecia subepidermal in origin, erumpent, with peridium, typically roestelioid or rarely aecidioid; spores catenulate, verrucose, most with pigmented wall. Uredinia subepidermal in origin, erumpent, lacking in most species; spores borne singly on pedicels, verrucose, pores scattered. Telia subepidermal in origin, erumpent as cushions, crests, or horns, gelatinizing when moist, mostly perennial in the host, often inducing swelling or fasciations; spores 1- to several-celled but mostly 2-celled by horizontal septum, borne singly on gelatinizing pedicels (with perhaps rare exceptions), germ pores 1 to several but most often 2 per cell, variously placed, wall mostly pigmented and smooth, germination occurs without dormancy, basidium external.

LECTOTYPE: Gymnosporangium fuscum DC.

Gymnosporangium, with 57 species (Kern) is a genus of predominantly north temperate climates. Most species are heteroecious and demicyclic. For anomalous urediniospore-teliospore situations see Hiratsuka (1973) for G. gaeumannii ssp. albertensis. Gymnosporangium is unusual because the telial state is on gymnosperms and the aecial state on dicot plants, especially the Pomoideae of the Rosaceae. There are 4 species, all microcyclic, all Asiatic and all on Rosaceae that have been described as species of Coleopuccinia or Coleopucciniella. It is logical to assume that they are derivatives of demicyclic species of Gymnosporangium, in accordance with Tranzschel's Law. Gelatinization of the telia occurs as in other species.

The variation among species makes a precise generic description nearly impossible. Type 4 spermogonia occur in all species where the lifecycle is known and gelatinizing teliospore pedicels characterize the majority of the species.

Economically important species include G. juniperi-virginianae, G. clavipes, G. fuscum, and G. yamadae.

REFERENCES: Bernaux, P. 1956. Contribution a l'étude de la biologie des Gymnosporangium. Ann. Epiphyt. 7:1-210. Crowell, I.H., 1940 The geographical distribution of the genus Gymnosporangium. Can. J. Res. C. 18:469-488. Hiratsuka, N. 1936, 1937. Gymnosporangium of Japan. Bot. Mag. Tokyo 50:481-488; 549-555; 593-599; 661-668; 51:1-8. Hiratsuka, Y. 1973. Sorus development, spore morphology, and nuclear condition of Gymnosporangium gaeumannii ssp. albertensis. Mycologia 65: 137-144. Kern, F.D., 1911. A biologic and taxonomic study of the genus Gymnosporangium. Bull. N.Y. Bot. Gard. 7:391-483. Kern, F.D. 1973. A revised account of Gymnosporangium. Penn. State Univ. Press. 134 pp. Laundon, G.F. 1975. Taxonomy and nomenclature notes on Uredinales. Mycotaxon 3:133-160. Parmelee, J.A. 1965. The genus Gymnosporangium in Eastern Canada. Can. J. Bot. 43:239-267. Parmelee, J.A. 1971. The genus Gymnosporangium in Western Canada. Can. J. Bot. 49: 903-926. Parmelee, J.A. and Corlett, M. 1978. Development of the aecium and the nuclear condition of Roestelia brucensis. Rept. Tottori Mycol. Inst. (Japan) 10:189-201. Peterson, R.S. 1967. Studies of juniper rusts in the West. Madroño 19:79-91. Savile, D.B.O. 1979. Fungi as aids in higher plant classification. Bot. Rev. 45:377-503. Thirumalachar, M.J. and Whitehead, M.D. 1954. On the validity of the genera Coleopuccinia and Coleopucciniella (Uredinales). Amer. J. Bot. 41:120-122.

Illustrations next page.

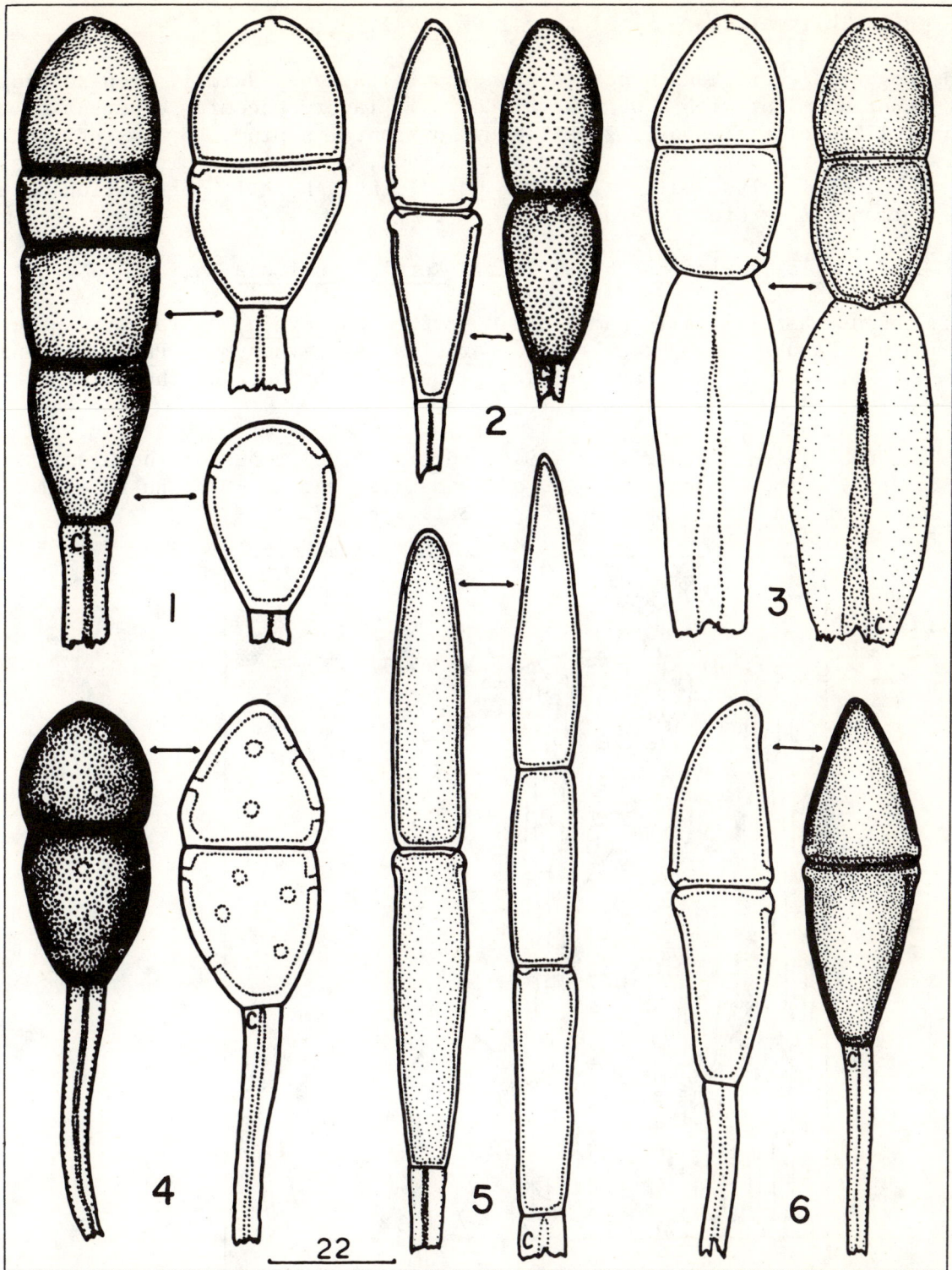

Gymnosporangium species; teliospores. 1, _libocedri_ (P. Henn.) Kern.
2, _clavariaeforme_ (Pers.) DC. 3, _clavipes_ (Cke. & Peck) Cke. & Peck.
4, _multiporum_ Kern. 5, _ellisii_ (Berk.) Ellis. 6. juniperi-virinianae
Schw.

60   KERNELLA Thirumalachar, Mycologia 41:97. 1949.

Spermogonia, aecia, and uredinia unknown. Telia subepidermal, deep-seated, without a peridium but with a peripheral "dermal" layer, becoming erumpent as long, hair-like columns; spores 2-celled by horizontal septum, borne singly on pedicels, new spores push older spores outward and, since the spores or pedicels adhere, a column results, wall pale, germ pore 1 in each cell, germination occurs without dormancy, basidium external.

TYPE: Kernella lauricola (Thir.) Thir. (Kernia lauricola Thir.).

The fungus was originally named and described as Kernia, a preempted name. Ragunathan and Ramakrishnan have treated Kernella as a synonym of Puccinia, and this is not illogical, but we are retaining the genus because of the columnar telia. The rust is on Litsea (Lauraceae).

REFERENCES: Thirumalachar, M.J. 1946. Kernia, a new genus of the Uredinales. Mycologia 38:679-686. Ragunathan, A.N. and Ramakrishnan, K. 1973. Rust fungi of Madras State. VI. Puccinia. Mysore J. Agr. Sci. 7:62-72.

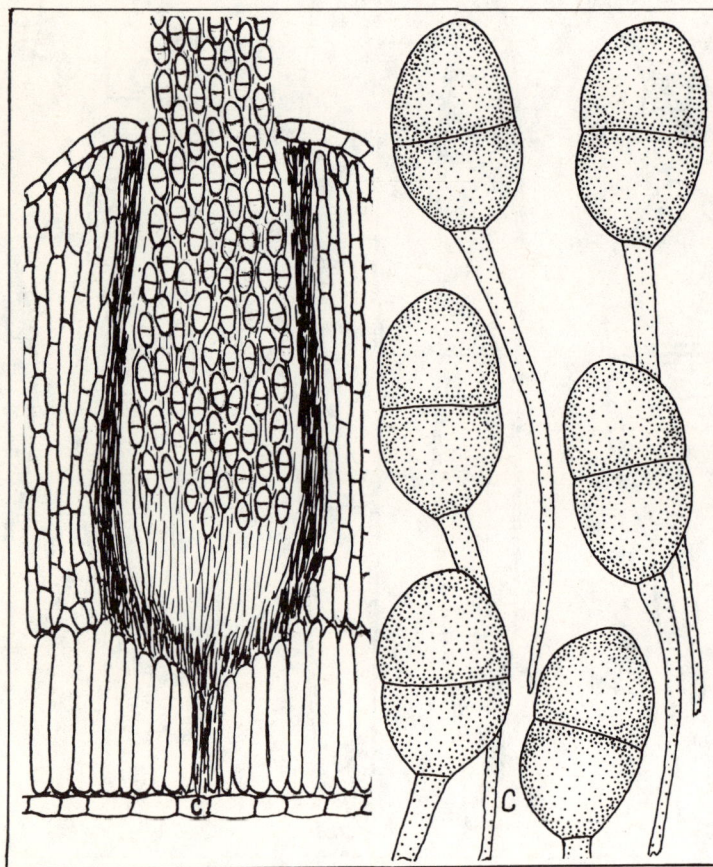

K. lauricola (Thir.) Thir.; sorus and telio-spores.

61. DIDYMOPSORELLA Thirumalachar, Sydowia 5:28. 1951.

Spermogonia subepidermal, type 5. Aecia subepidermal in origin, erumpent, uredinoid; spores borne singly on pedicels, echinulate. Uredinia subepidermal in origin, erumpent; spores as the aeciospores. Telia subepidermal in origin, with a peripheral layer of hyphae and host cells, erumpent as loosely adherent filaments of spores and intermixed gelatinizing paraphyses; spores 2-celled, borne singly on pedicels, germ pore 1 or 2 in each cell, wall pale, germination occurs without dormancy, basidium external.

TYPE: Didymopsorella toddaliae (Thir. & Mund.) Thir. (= D. lemanensis Doidge) Hirat. f.).

Two species have been recognized. The paraphyses usually have been overlooked but were described and illustrated by Ramakrishnan when he described the synonymous genus Gymnopuccinia. Both species are on Toddalia (Rutaceae).

REFERENCE: Ramakrishnan, T.S. 1951. Two new rusts from South India. Trans. Brit. Mycol. Soc. 34:141-145.

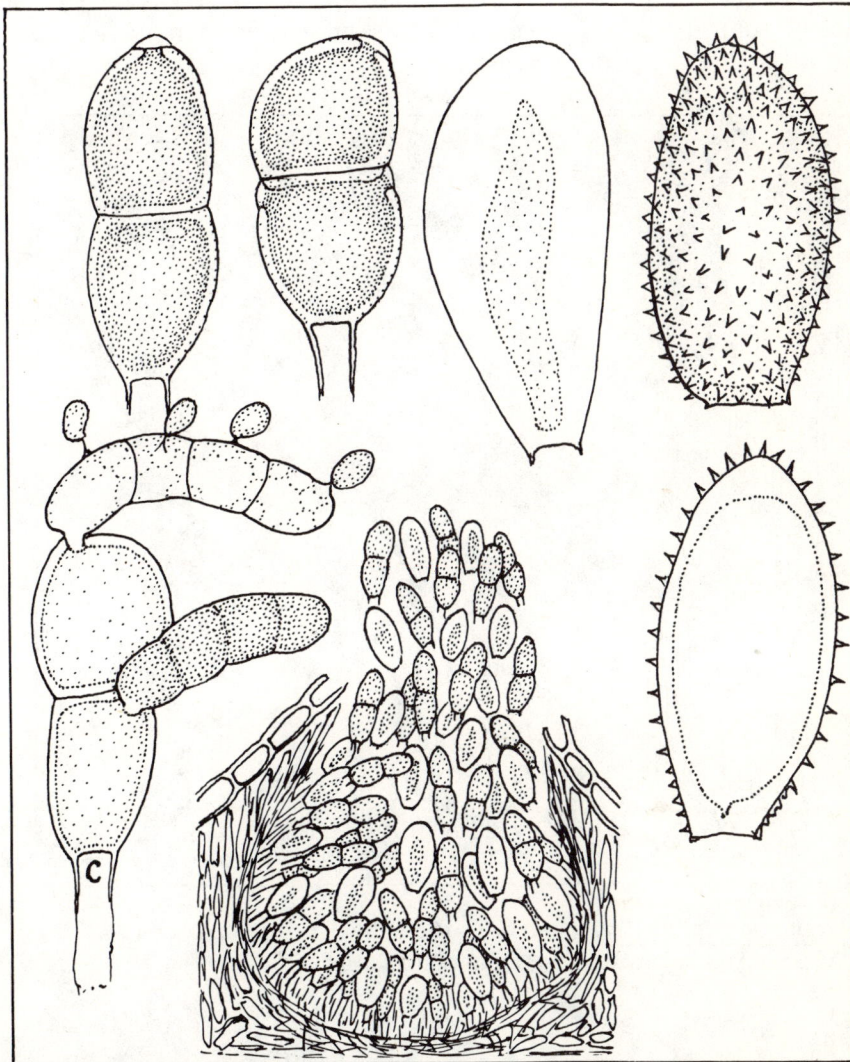

D. lemanensis (Doidge) Hirat. f.; sorus, teliospores, one paraphysis, and urediniospores.

62. MIYAGIA Miyabe in H. & P. Sydow, Ann. Mycol. 11:107. 1913.

Spermogonia subepidermal, type 4. Aecia subepidermal in origin, erumpent, uredinoid, with peridium as in the uredinia; spores borne singly on pedicels, echinulate, pores obscure. Uredinia subepidermal in origin, erumpent, with a peridium of laterally adherent, palisade-like paraphyses; spores borne singly on pedicels, echinulate. Telia subepidermal in origin, tardily exposed, with a peridium as in the uredinia; spores 2-celled by horizontal septum, borne singly on pedicels, wall pigmented, germ pore 1 in each cell, basidium external.

TYPE: Miyagia pseudosphaeria (Mont.) Joerst. (M. anaphalidis Miyabe).

Miyagia differs from Puccinia only in the peridiate uredinia and telia. There are three species. In the type species 1-celled teliospores are sometimes commoner than 2-celled spores. Also on Compositae is Corbulopsora with similarly peridiate uredinia and telia but with only 1-celled teliospores and with aecidioid aecia. The relationship of Miyagia doubtless is with Puccinia and that of Corbulopsora with Uromyces and in the 1959 Illustrated Genera they were treated as synonyms.

REFERENCES: Hiratsuka, N. 1969. Notes on the genus Miyagia Miyabe ex Syd. Trans. Mycol. Soc. Japan 10:89-90. Wilson, M. and Henderson, D.M. 1966. British rust fungi. Cambridge Univ. Press. 384 pp.

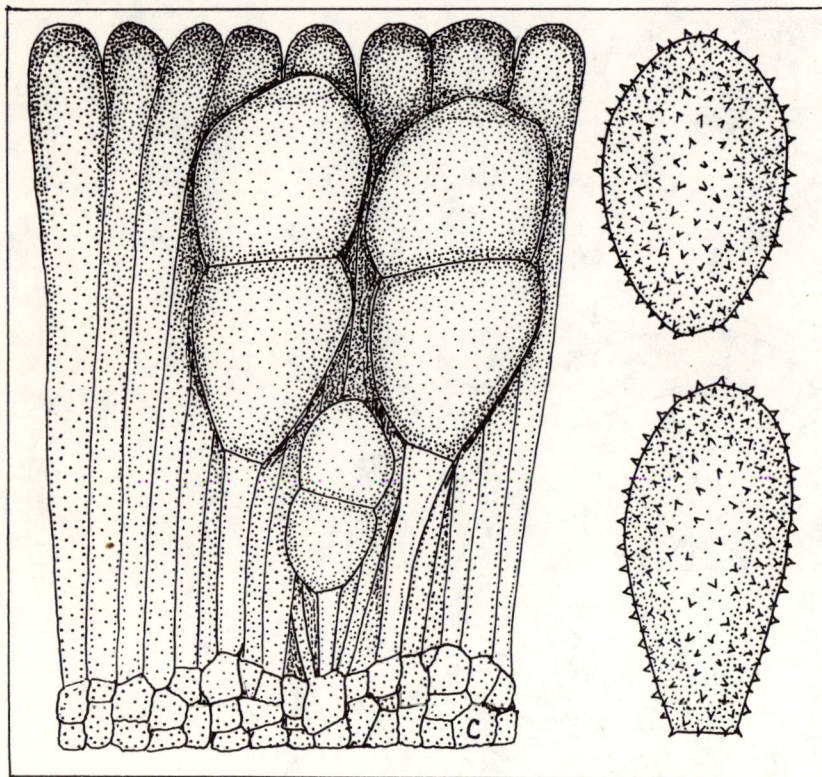

M. pseudosphaeria (Mont.) Joerst.; peridial cells, teliospores, and urediniospores.

63. PUCCINIA Persoon, Syn. Meth. Fung. p. 225. 1801.

Spermogonia subepidermal, type 4. Aecia subepidermal in origin, erumpent, aecidioid with peridium and catenulate, mostly verrucose spores, or uredinoid without peridium and with mostly echinulate spores borne singly on pedicles (in this case sometimes called primary uredinia). Uredinia subepidermal in origin, erumpent, without peridium but may have paraphyses; spores borne singly on pericels, mostly echinulate, pores various. Telia subepidermal in origin, erumpent in most species but remaining covered by the epidermis and sometimes divided into locules by stromatic paraphyses in some species; spores typically 2-celled by horizontal septum (but may have 1-celled and sometimes 3- or 4-celled spores in some species) borne singly on pedicels, basal sporogenous cells from none to highly developed, spore wall mostly pigmented but degree of pigmentation varies widely, smooth or variously sculptured, germ pore 1 in each cell but not differentiated in a few species, basidium external

LECTOTYPE: *Puccinia graminis* Pers. ex Pers.

Puccinia is much the largest genus of Uredinales with some 3000 to 4000 species. Heteroecious and autoecious species occur and also life cycle variations down to the ultimate of teliospores only. The Compositae, Cyperaceae, Gramineae, and Liliaceae serve as hosts for large numbers of species but most groups of vascular plants are parasitized. There are species in all land areas except the polar regions.

Many serious diseases are caused by species of *Puccinia*, e.g. black stem rust (*P. graminis*) of small grains and grasses, leaf or brown rust (*P. recondita*) of small grains and grasses, stripe or yellow rust (*P. striiformis*) of wheat and grasses, crown rust (*P. coronata*) of oats and grasses, sorghum rust (*P. purpurea*), common corn rust (*P. sorghi*), sugar cane rusts (*P. kuehnii*, *P. melanocephala*), sunflower rust (*P. helianthi*), safflower rust (*P. calcitrapae* var. *centaureae*), cotton rust (*P. cacabata*), asparagus rust (*P. asparagi*), mint rust (*P. menthae*), snapdragon rust (*P. antirrhini*), hollyhock rust (*P. malvacearum*), and others.

REFERENCES: Arthur, J.C. 1907-1927. Uredinales. N. Amer. Flora 7:83-848 (but with *Puccinia* broken into several "life cycle" genera); Azbukina, Z.M. 1974. Rust fungi of the Soviet Far East (in Russian). Akad. Nauk. Moscow. 257 pp; Cummins, G.B. 1971. The rust fungi of cereals, grasses and bamboos. Springer-Verlag, New York. 570 pp.; Cummins, G.B. 1978. Rust fungi on legumes and composites in North America. Univ. of Arizona Press. 424 pp.; Gaeumann, E. 1959. Die Rostpilze Mittleuropas. Buechler & Co., Bern. 1407 pp.; León-Gallegos, H.M. and Cummins, G.B. 1981. Uredinales (Royas) de México. Sect. Agric y Recursos Hidrol. México. Vol. I, 438 pp. Lindquist, J.C. 1982. Royas de la Republica Argentina y zonas limitrofes. Coleccion Cient. XX:213-476. Majewski, T. 1979. Grzyby (Mycota). Tom. XI. Uredinales II. Polska Akad. Nauk. Warszawa-Krakow. 462 pp. Ul'yanischev, V.I. 1978. Key to the rust fungi of the USSR. Vol. II (in Russian). Akad. Nauk. 383 pp. (322 pp. devoted to *Puccinia*). Numerous other descriptive manuals.

Illustrations next page.

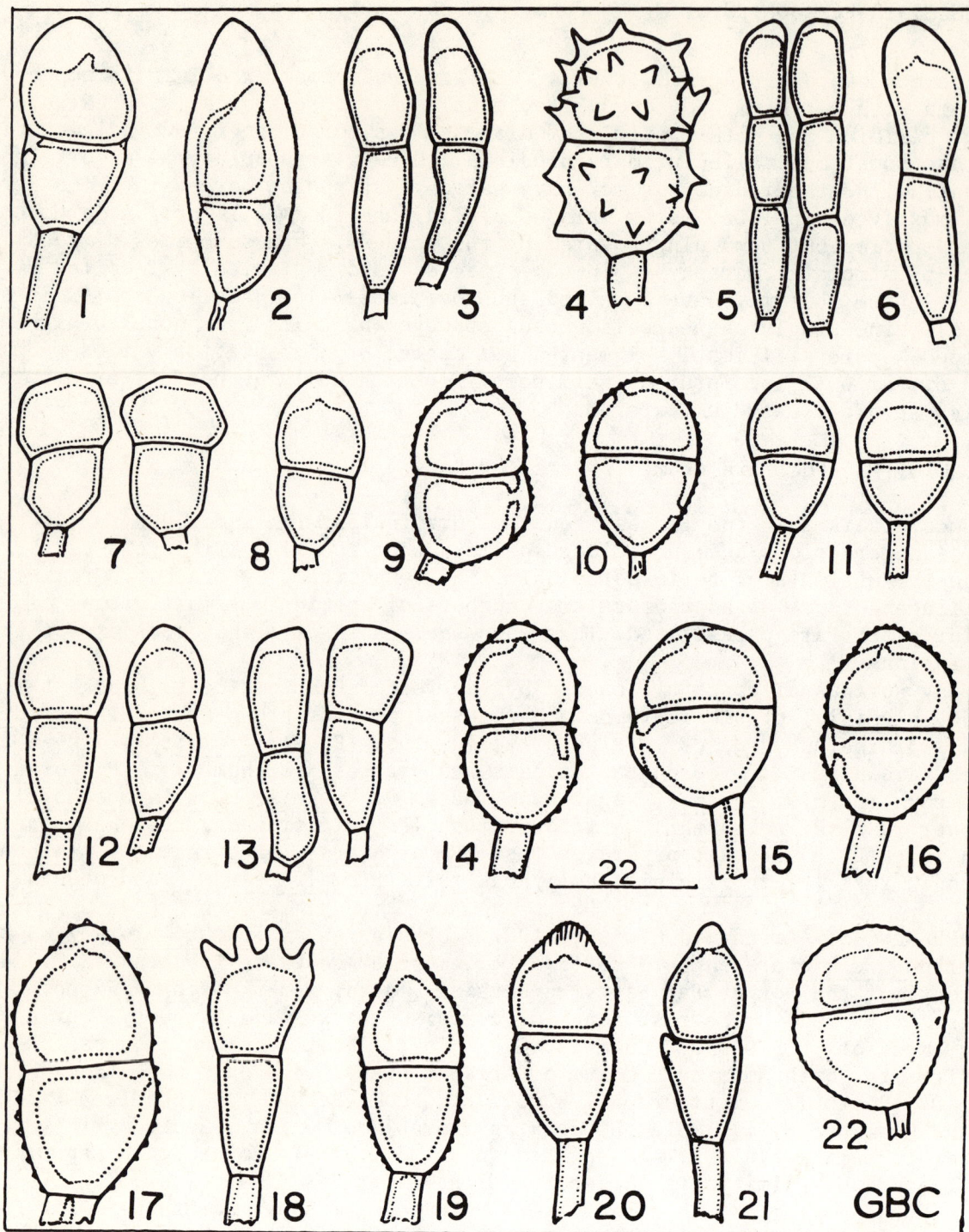

Puccinia species; teliospores. 1, spegazziniana DeT. 2, corylopsidis Cumm. 3, scirpi-ternatani Hirat. f. 4, prostii Duby. 5, scleriae (Paz.) Arth. 6, heliconiae Arth. 7, polysora Underw. 8, sorghi Schw. 9, hogsoniana Kern. 10, cyani Pass. 11, cacabata Arth. & Holw. 12, melanocephala Syd. 13, recondita Rob. 14, conoclinii Seym. 15, solidipes Jack. & Holw. 16, anisacanthi Diet. & Holw. 17, batatae Syd. 18, coronata Cda. 19, longi-pedicellata Barth. 20, tanaceti DC. 21, arachidis Speg. 22, wolgensis Nawash.

63E.   ENDOPHYLLUM Léveillé, Mém. Soc. Linn. Paris 4:208. 1825.

   Spermogonia subepidermal, type 4 or wanting. Aecia and uredinia not produced. Telia with a peridium, resembling Aecidium; spores 1-celled, catenulate, like
aeciospores except producing a basidium when germinating.

   TYPE: Endophyllum sempervivi (Alb. & Schw.) DeBary (E. persoonii Lév.).

   Endophyllum is composed of microcyclic species that simulate the anamorph
genus Aecidium. In theory, the species may be derived from any macrocyclic or
demicyclic rust fungus that has aecia of the Aecidium type and type 4 spermogonia.
Monosporidium (Kulkarniella) differs because of type 7 (or 5?) spermogonia. The
kind of spermogonium could prove useful in indicating a possible parental long
cycle genus. The endo forms are really life cycle variants of macrocyclic species
in the same sense as are species of "Micropuccinia." Such species are not of Aecidium either because the spores are teliospores. The endo genera are sort of pseudo-
genera.

   REFERENCE: Buriticá, P. and Hennen, J.F. 1980. Pucciniosireae (Uredinales,
Pucciniaceae) Flora Neotropica, Monogr. 24:10-14.

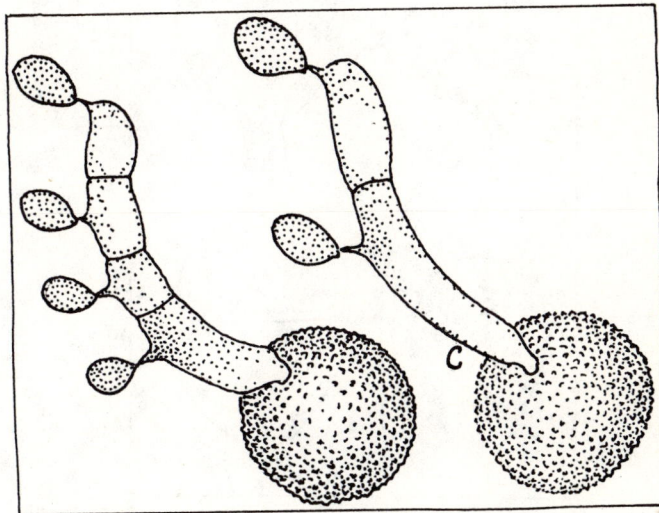

Endophyllum; teliospores; schematic.

64.   STEREOSTRATUM Magnus, Ber. Dtsch. Bot. Ges. 17:181. 1899.

    Spermogonia and aecia unknown. Uredinia subepidermal in origin, erumpent;
spores borne singly on pedicels, echinulate, pores equatorial. Telia subepidermal
becoming erumpent in large Stereum-like crusts; spores 2-celled by horizontal septum, borne singly on pedicels, germ pores 3 per cell, obscure, basidium presumably
external.

    TYPE: Stereostratum corticioides (Berk. & Br.) Magn. (Puccinia corticioides
Berk. & Br.).

    Only one species is known. It parasitizes various genera of the Bambusoideae
in China, Japan, and probably elsewhere. According to Thirumalachar (Mycologia
39:334-340. 1947) the teliospores are produced in succession with older spores
being pushed out laterally, thus producing a spreading, cushion-like sorus.

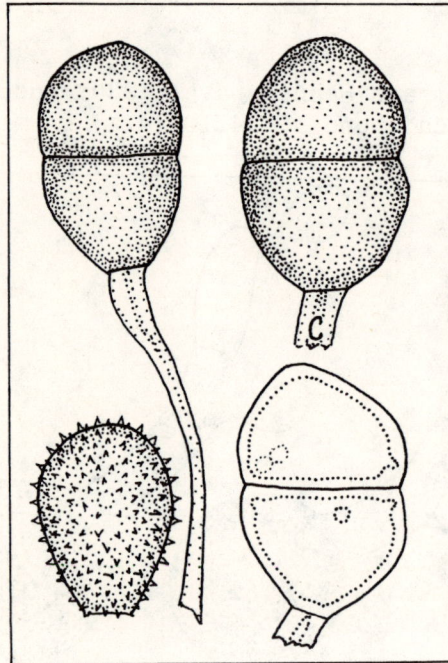

S. corticioides (Berk. & Br.)
Magn.; one urediniospore and
teliospores.

65. CHRYSOCYCLUS H. Sydow, Ann. Mycol. 23:322. 1925.

Spermogonia subepidermal, type 4. Aecia and uredinia unknown, probably not produced. Telia subepidermal in origin, erumpent; spores (or basidia?) 2-celled, borne singly on pedicels, each cell becoming 4-celled by apical growth and by the subsequent division of the protoplast into 4 parts each of which produces a sterigma and basidiospore, wall pale, germination occurs without dormancy.

LECTOTYPE: Chrysocylclus cestri (Diet. & P. Henn.) Syd. (Puccinia cestri Diet. & P. Henn.).

There are three species and a possible fourth known in South and Central America. Jackson (Mycologia 24:62-186 (80). 1932) suggests that the genus is little more than an exaggerated Puccinia. The basidium of the lower cell is not in the teliospore cell but that of the upper cell comes close to being an internal basidium. Perhaps the genus is midway between Puccinia and Chrysopsora and both may be derivatives of Puccinia. Buriticá treats Chrysella as a synonym.

REFERENCE: Buriticá, p. 1974. A revision of rust genera (Uredinales) with reduced life cycles. (Ph.D. thesis, Purdue University. 114 pp.).

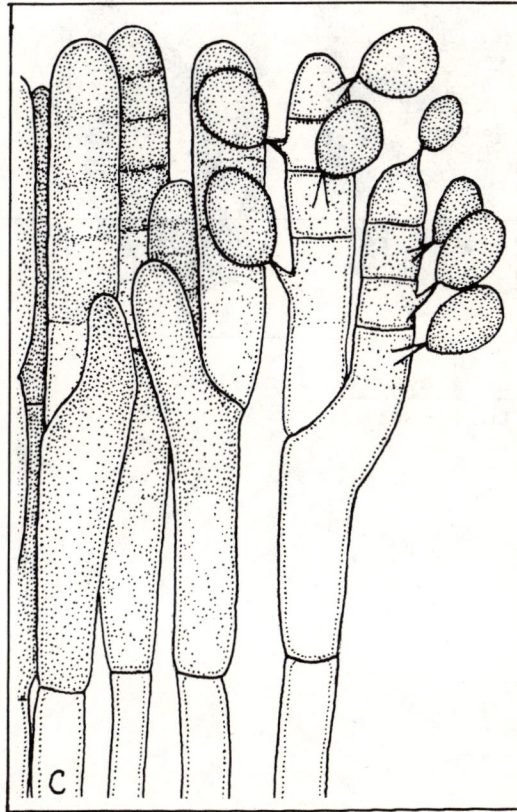

C. cestri (Diet. & P. Henn.) Syd.; teliospores and basidiospores.

66.   DASYSPORA Berkeley & Curtis, J. Philadel. Acad. Sci. ser. 2; 2:281. 1853.

Spermogonia subepidermal, type 5. Aecia and uredinia unknown, probably not produced. Telia subepidermal in origin, erumpent; spores 2-celled by horizontal septum, borne singly on pedicels, wall pigmented, beset with rod-like verrucae mostly branched apically, germ pore 1 in each cell, basidium external.

TYPE: Dasyspora gregaria (Kunze) P. Henn. (D. foveolata Berk. & Curt.).

The type is the only species and it occurs in the American tropics on Anonaceae. Dasyspora differs from Puccinia in the type of spermogonium. The reported "hyphoid" aecium is an alga (Hennen and Figueiredo).

REFERENCE: Hennen, J.F. and Figueredo, M.B. 1981. The hyphoid aecium, a rust-alga association (Dasyspora-Stomatochroon), and other corrections to neotropical rusts (Uredinales). Mycologia 73:350-355.

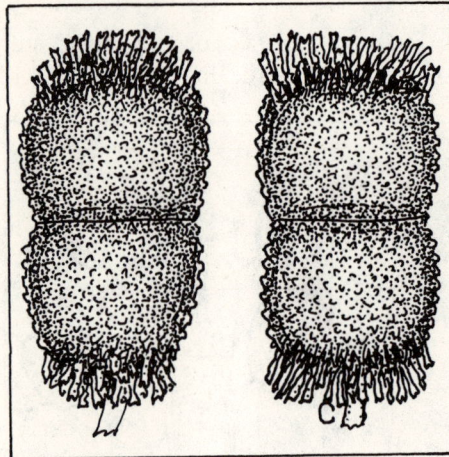

D. gregaria (Kunze) P. Henn.;
teliospores.

67.   SORATAEA H. Sydow, Ann. Mycol. 28:432.1930.

Spermogonia subcuticular, type 7. Aecia subepidermal in origin, erumpent, either uredinoid with spores borne singly or aecidioid with catenulate spores and a peridium. Uredinia subepidermal in origin, erumpent, with or without paraphyses; spores borne singly on pedicels, echinulate, pores equatorial. Telia subepidermal in origin, erumpent; spores 2- to several-celled by horizontal septa, borne singly on pedicels grouped on basal sporogenous cells in at least some species, wall pale but usually pigmented, pore obscure if differentiated, apparently mostly germinating by growth of the apex of the cells, without dormancy, basidium external.

TYPE: Sorataea amiciae H. Syd.

Only S. amiciae (Allopuccinia diluta H.S. Jack. & Holw.) was recognized in the 1959 Illustrated Genera but it was suggested that other species described in Puccinia might belong. Savile transferred Puccinia baphiae because of type 7 spermogonia and Eboh and Cummins transferred four more species because of the spermogonia. We now transfer Mimema holwayi H.S. Jack. as S. holwayi (H.S. Jack.) comb. nov. In 1948, Thirumalachar and Cummins suggested that Leucotelium Tranz. did not differ in any substantial way. The species of Sorataea mentioned above occur on legumes and are autoecious. The species of Leucotelium are heteroecious with aecial states on Ranunculaceae and telial states on Rosaceae. Insofar as known, the species have type 7 spermogonia. We believe that Leucotelium should be treated as a synonym of Sorataea and so propose the following transfers: Sorataea cerasi (Cast.) comb. nov. (Puccinia cerasi Cast., Leucotelium cerasi Tranz.), S. pruni-persicae (Hori) comb. nov. (Puccinia pruni-persicae Hori, Leucotelium pruni-persicae Tranz.), and S. padi (Tranz.) comb. nov. (Leucotelium padi Tranz.).

REFERENCES: Eboh, D.O. and Cummins, G.B. 1980. Species of Sorataea (Uredinales). Mycologia 72:203-204. Savile, D.B.O. 1971. Generic disposition and pycnium type in Uredinales. Mycologia 63:1089-1091. Thirumalachar, M.J. and Cummins, G.B. 1948. Status of Allopuccinia, Leucotelium, Edythea, and Ypsilospora. Mycologia 40:417-422.

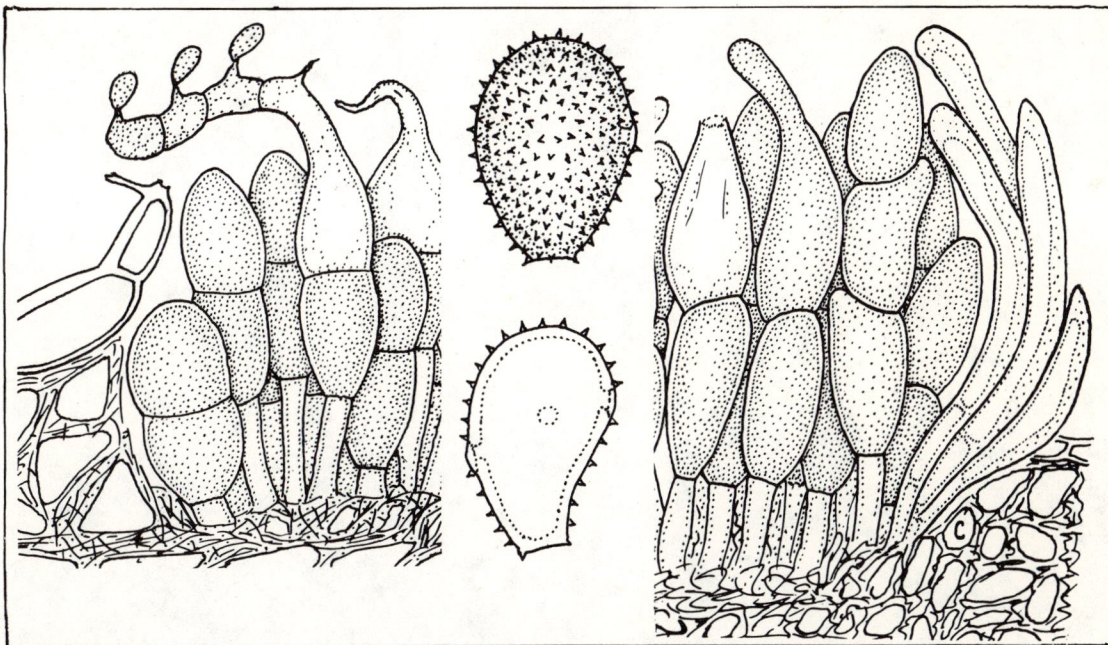

S. cerasi (Cast.) Cumm. & Y. Hirat.; teliospores (left). S. amiciae Syd.; urediniospores and a telium (right).

68. POROTENUS Viégas, Bragantia 19(pt. 1): XCVIII. 1960.

Spermogonia type 5 or 7. Aecia subepidermal in origin, erumpent, uredinoid; spores borne singly on pedicels, echinulate, pores equatorial. Uredinia subepidermal in origin, erumpent; spores as the aeciospores. Telia subepidermal in origin, erumpent; spores 2-celled by horizontal septum, borne singly on pedicels, each cell germinating by elongation of the apex , with a differentiated area in the wall, or with a discrete germ pore.

TYPE: Porotenus concavus Viégas.

The type species may attack leaf blades and stems, often causing hypertrophy and distortion. The urediniospores are radially asymmetrical. The species occurs in Brazil on Memora of the Bignoniaceae. Three species, previously assigned first to Puccinia and then to Prospodium are here transferred to Porotenus: P. elatipes (Arth. & Holw.) comb. nov. (Puccinia elatipes Arth. & Holw.), P. permagnum Arth. & Holw.) comb. nov. (Puccinia permagna Arth. & Holw.), and P. depallens (Arth. & Holw.) comb. nov. (Puccinia depallens Arth. & Holw.). Two of the species are macrocyclic and two are microcyclic.

Porotenus has characters in common with Prospodium, where three species were placed by Cummins, but the teliospores were misfits there. P. concavus is scarcely separable from Sorataea. The present association may or may not be an improvement.

REFERENCE: Cummins, G.B. 1940. The genus Prospodium (Uredinales). Lloydia 3:1-78.

Illustrations next page.

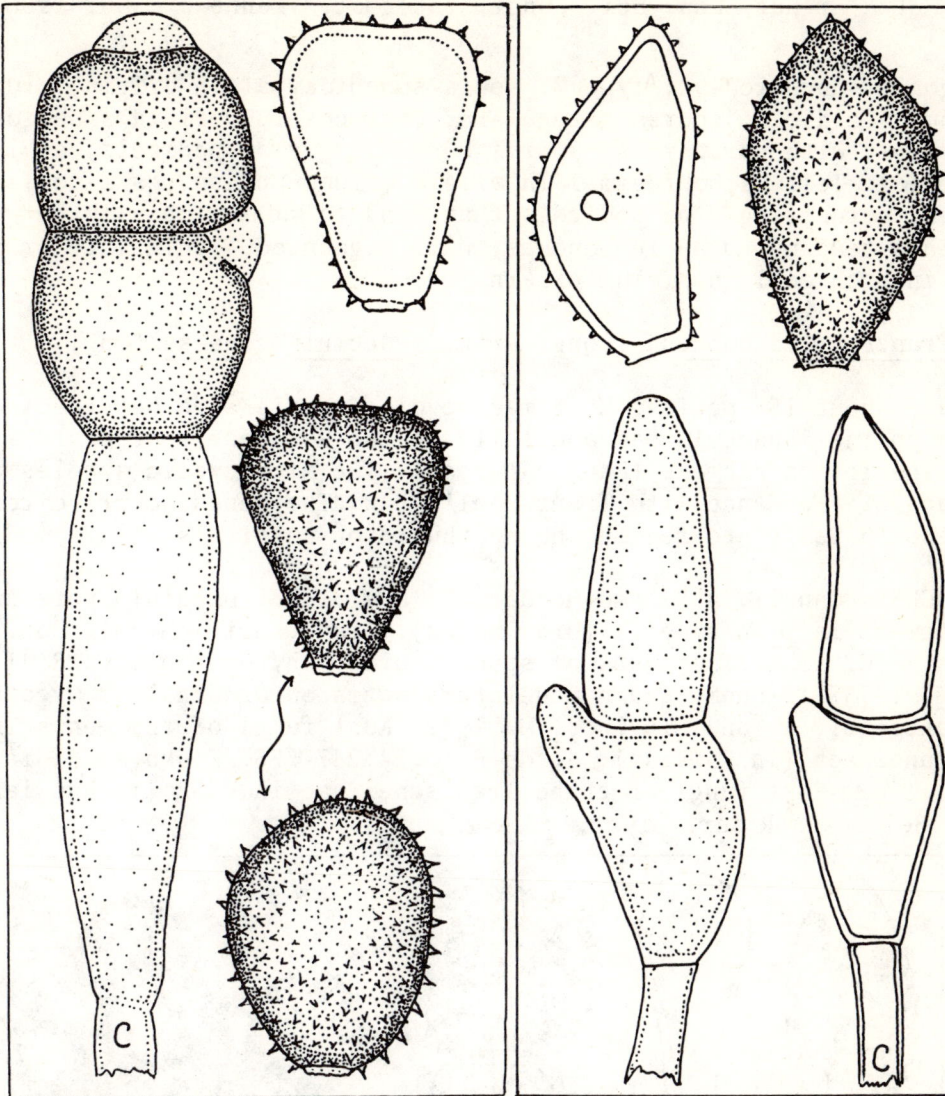

P. elatipes (Arth. & Holw.) Cumm. & Y. Hirat.; teliospores
and urediniospores (left). P. concavus Viegás; urediniospores
and teliospores (right).

69. TRANZSCHELIA Arthur, Rés. Sci. Congr. Internat. Vienne p. 340. 1906

Spermogonia subcuticular, type 7. Aecia subepidermal in origin, erumpent, with peridium, aecidioid; spores catenulate, verrucose. Uredinia subepidermal in origin, erumpent, with paraphyses; spores borne singly on pedicels, echinulate, pores equatorial. Telia subepidermal in origin, erumpent; spores 2-celled by horizontal septum, borne singly on pedicels that tend to adhere basally in groups, the the spores thus tending to be in bunches, wall pigmented, echinulate or verrucose, germ pore 1 in each cell, basidium external.

TYPE: Tranzschelia cohaesa (Long) Arth. (Puccinia cohaesa Long).

There are about 15 species. Most macrocyclic species are heteroecious, producing aecia on the Ranunculaceae and telia on the Rosaceae, but there is one macrocyclic species on Anemone (Ranunculaceae). The microcyclic species are on Ranunculaceae in accordance with Tranzschel's Law. The genus occurs circumglobally in the north temperate region and southward on cultivars.

REFERENCES: Bennell, A.P. and Henderson, D.M. 1978. Urediniospore and teliospore development in Tranzschelia (Uredinales). Trans. Brit. Mycol. Soc. 71: 271-278. Dunegan, J.C. 1938. The rusts of stone fruits. Phytopathology 28:411-427. Laundon, G.F. 1975. Taxonomy and nomenclature notes on Uredinales. Mycotaxon 3: 133-161. Tranzschel, W. and Litvinov, M. 1939. Rust fungi of the genus Tranzschelia on the Prunoideen (in Russian). J. de Bot. 24:247-253. Zenkova, E. Ya and Kuprevich, V.F. 1962. Rust fungi of tribe Tranzschelieae in USSR (in Russian). Vesti Akad. Navuk Belarus SSR ser. biyal. 4:15-25.

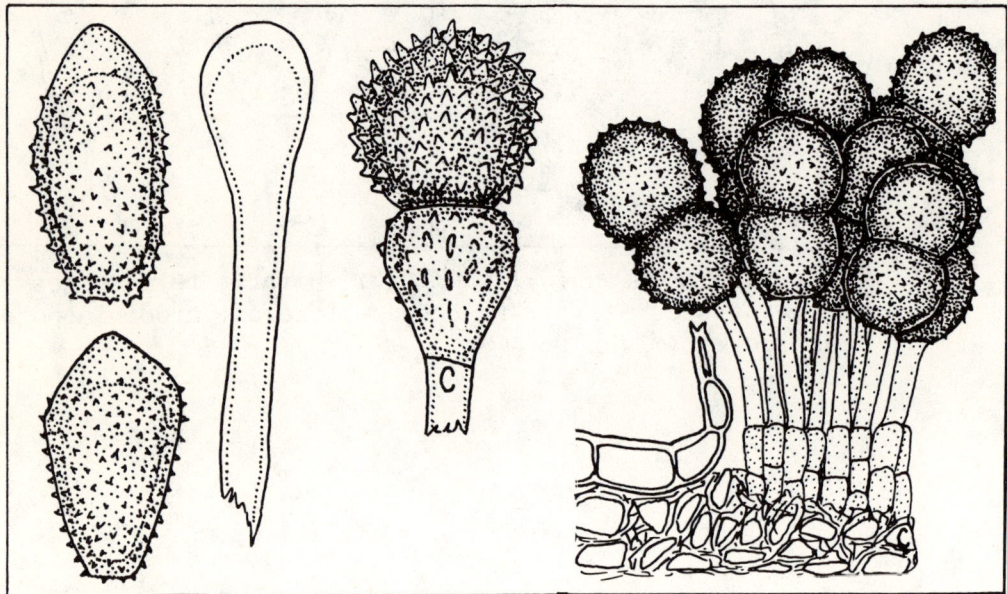

T. discolor (Fckl.) Tranz. & Litv.; urediniospores, one paraphysis and one teliospore (left). T. pruni-spinosae (Pers.) Diet.; teliospores (right).

70. PROSPODIUM Arthur, J. Mycol. 13:31. 1907.

Spermogonia subcuticular, type 7. Aecia subepidermal in origin, erumpent, uredinoid; spores borne singly on pedicels, mostly echinulate. Uredinia subepidermal in origin, erumpent in some species or substomatal in origin but emerging through stomata to sporulate above the leaf surface, i. e. suprastomatally, usually with peripheral paraphyses, the suprastomatal species basket-like with paraphyses on the rim; spores borne singly on pedicels, mostly echinulate, some radially asymmetrical, the wall simple or with an outer pale, hygroscopic layer, pores equatorial. Telia as the uredinia of the species; spores 2-celled by horizontal septum, borne singly on pedicels which often are appendaged basally, spore wall pigmented, often obviously bilaminate and usually echinulate or verrucose, germ pore 1 in each cell, basidium external.

TYPE: Prospodium appendiculatum (Wint.) Arth. (Puccinia appendiculata Wint.).

Some 50 species of Prospodium have been described; all are autoecious; some are microcyclic; most occur on Bignoniaceae and the remainder on Verbenaceae. The uredinoid aecia are never suprastomatal although the uredinia of the same species may be. Microcyclic species never produce suprastomatal telia although the puttive parental species may; they always assume the aecial habit.
Three sections have been recognized (but not formalized as subgenera): Euprospodium with erumpent uredinia and telia, Nephlyctis which is comprised of microcyclic species only, and Cyathopsora with suprastomatal uredinia and telia. Despite such apparently divergent morphology as subepidermal versus suprastomatal sori the species obviously are closely related.

REFERENCE: Cummins, G.B. 1940. The genus Prospodium (Uredinales). Lloydia 3:1-78.

Illustrations next page

105

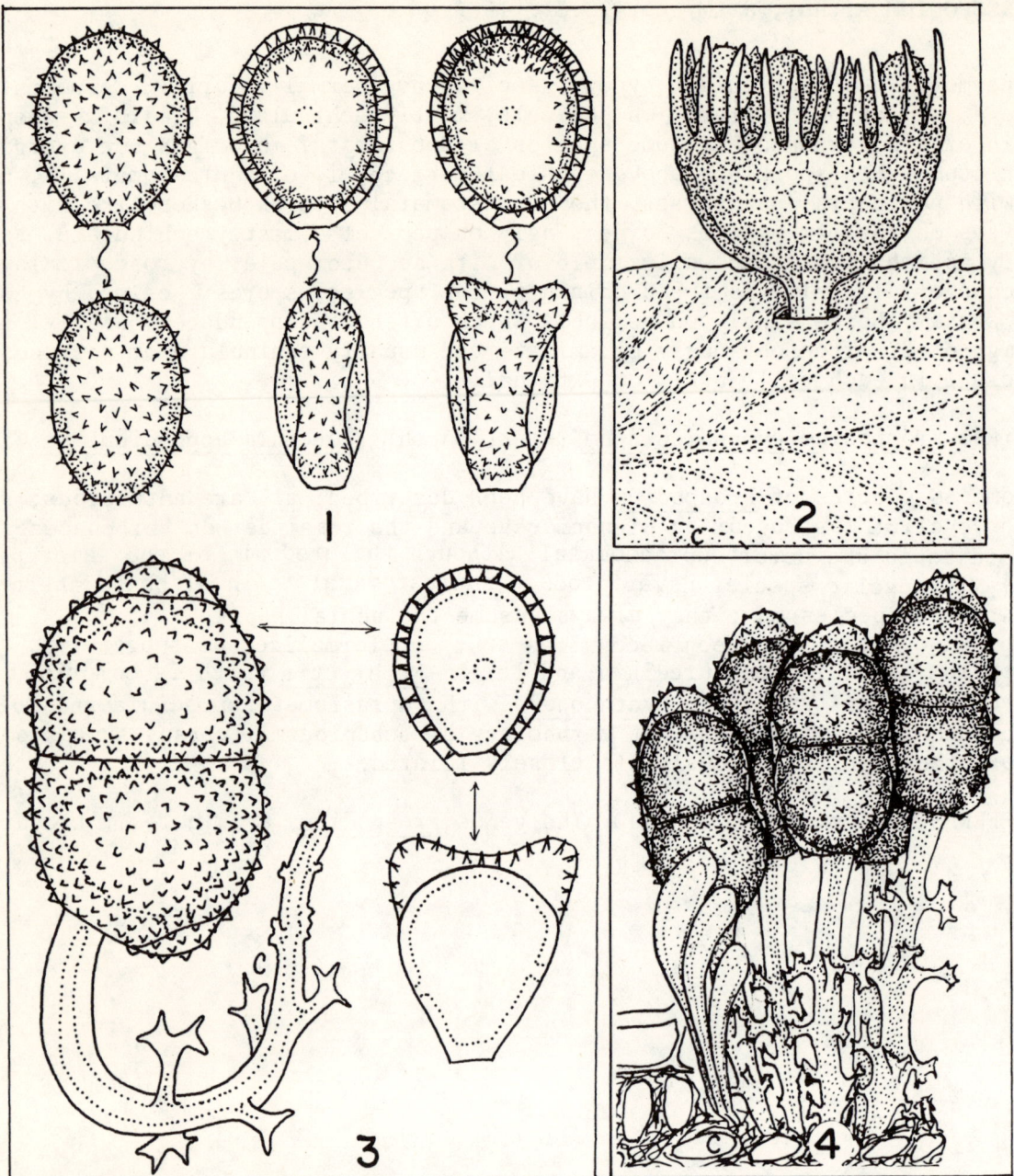

Prospodium. 1. Simple, unicapitate, and bicapitate urediniospores
(schematic). 2. Suprastomatal sorus (schematic). 3. P. couraliae Syd.;
one teliospore and urediniospores. 4. P. appendiculatum (Wint.) Arth.;
teliospores.

71. CUMMINSIELLA Arthur, Bull. Torrey Bot. Club 60:475. 1933.

Spermogonia subepidermal, type 4. Aecia subepidermal in origin, erumpent, with peridium, aecidioid; spores catenulate, verrucose. Uredinia subepidermal in origin, erumpent, with or without paraphyses; spores borne singly on pedicels, echinulate or verrucose, pores mostly zonate. Telia subepidermal in origin, erumpent; spores borne singly on pedicles, wall pigmented, often bilaminate, the surface variously sculptured, germ pores 2 in each cell, basidium external.

TYPE: Cumminsiella mirabilissima (Peck) Nannf. (Uropyxis sanguinea Arth.).

There are eight species, all on Mahonia-Berberis (Berberidaceae), all autoecious and macrocyclic, and all native to North and South America. C. mirabilissima has been introduced with its host into other regions. The aecial state was first recognized in Europe and has been collected only rarely in its original area. Cumminsiella differs from Puccinia because of the two pores per teliospore cell and from Uropyxis because of type 4 spermogonia and aecidioid aecia.

REFERENCES: Baxter, J.W. 1957. The genus Cumminsiella. Mycologia 49:864-873. McCain, J.W. and Hennen, J.F. 1982. Is the taxonomy of Berberis and Mahonia (Beberidaceae) supported by their rust pathogens, Cumminsiella santa sp. nov. and other Cumminsiella species (Uredinales)? Syst. Bot. 7:48-59.

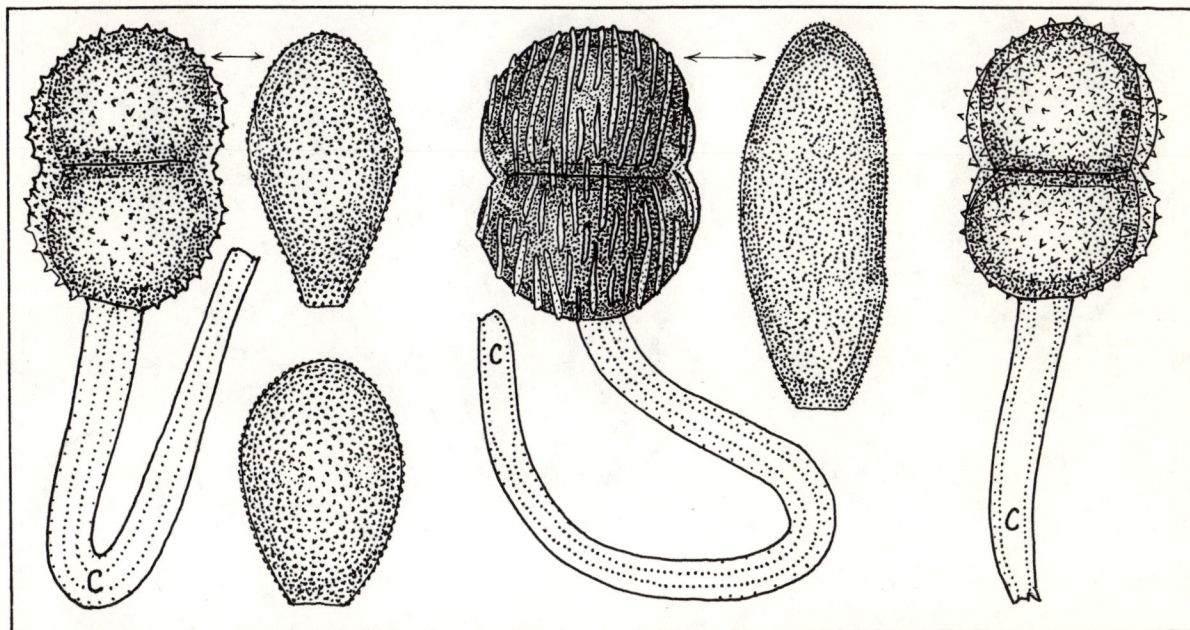

C. mirabilissima (Peck) Nannf.; one teliospore and two urediniospores (left).
C. wootoniana (Arth.) Arth.; one teliospore and one urediniospore (center).
C. standleyana Cumm.; one teliospore (right).

72. UROPYXIS Schroeter, Hedwigia 14:165. 1875.

Spermogonia subcuticular, type 7. Aecia subepidermal in origin, uredinoid; spores borne singly on pedicels. Uredinia subepidermal in origin, erumpent, with peripheral paraphyses; spores borne singly on pedicels, echinulate, pores scattered, obscure. Telia subepidermal in origin, erumpent; spores borne singly on pedicels which often are hygroscopic, 2-celled by horizontal septum, wall usually conspicuously hygroscopic and bilaminate, the outer wall often easily separable, sculptured or smooth, germ pores 2 in each cell, basidium external.

TYPE: Uropyxis amorphae (Curt.) Schroet. (Puccinia amorphae Curt.).

The species are commoner in warm regions but U. amorphae is known in Canada. the 15 or so species are autoecious; some are macrocyclic and some microcyclic. most occur on legumes. Uropyxis differs from Puccinia and Cumminsiella because of type 7 spermogonia and from Prospodium because of the two pores per teliospore cell.

REFERENCES: Baxter, J.W. 1959. A monograph of the genus Uropyxis. Mycologia 51:210-226. Cummins, G.B. 1978. Rust fungi on legumes and composites in North America. Univ. Arizona Press. 424 pp.

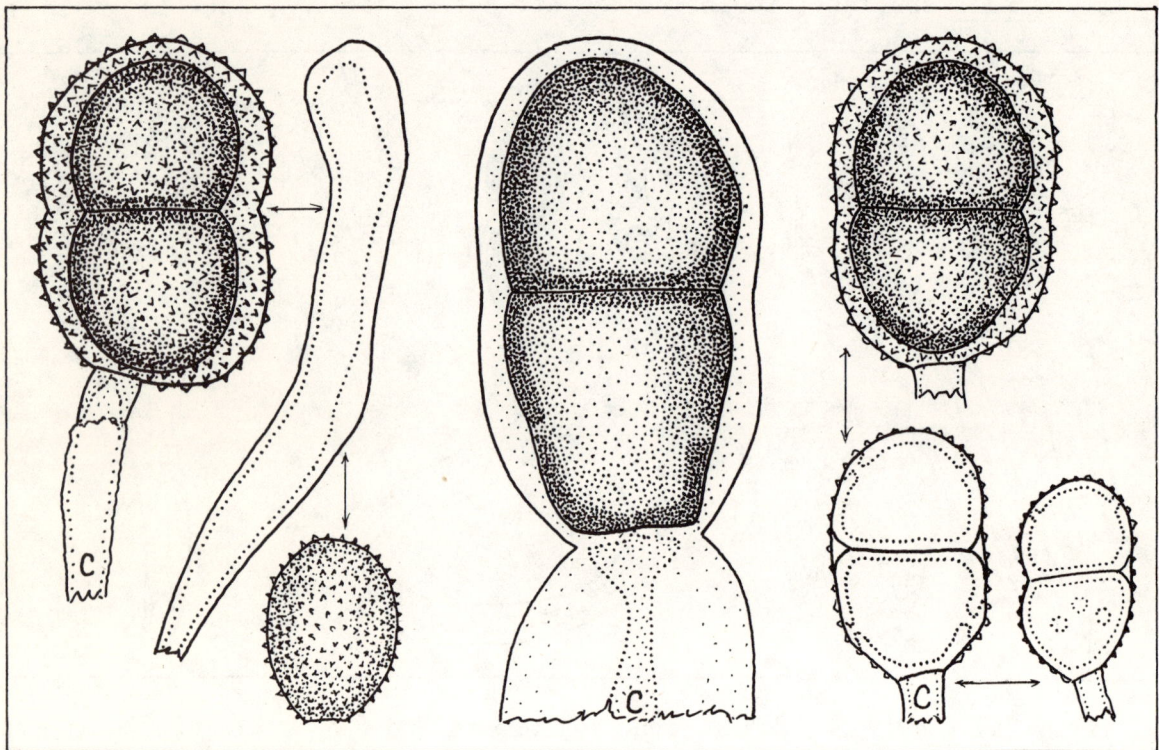

U. amorphae (Curt.) Schroet.; one teliospore, one paraphysis, and one urediniospore (left). U. holwayi (Arth.) Arth.; one teliospore (center). U. heterospora Hennen & Cumm.; teliospores (right).

73. CLEPTOMYCES Arthur, Bot. Gaz. 65:464. 1918.

Spermogonia subepidermal, type 4. Aecia and uredinia unknown. Telia subepidermal in origin, erumpent; spores 2-celled by horizontal septum, borne singly on pedicels, wall pigmented and bilaminate, germ pores 4 or more in each cell, germination not recorded but basidium undoubtedly external.

TYPE: Cleptomyces lagerheimianus (Diet.) Arth. (Puccinia lagerheimiana Diet.)

The type on Verbenaceae and one other species are known. Both occur in South America. The genus differs form Puccinia and Prospodium because of multiple germ pores and from Prospodium in the type of spermogonium. Stereostratum, with only telia known and parasitizing Bambusoideae, has 3 pores per cell. Thirumalachar (Mycologia 52:688-693. 1960) considered the genus to be synonymous to Cleptomyces and transferred C. lagerheimianus to Stereostratum. When the aecial states become known, the question of relationship may be decided.

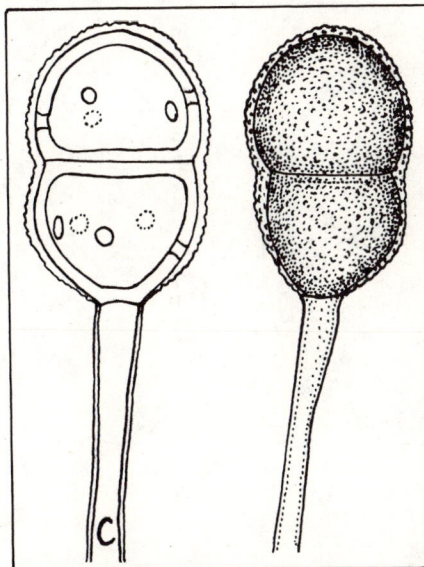

C. lagerheimianus (Diet.)
Arth.; teliospores.

74. MACRUROPYXIS Azbukina, Komarov Readings, Vladivostok 19:20-21. 1972.

Spermogonia subcuticular, type 7. Aecia and uredinia unknown, probably not produced. Telia subepidermal in origin, erumpent, without paraphyses; spores 2-celled by horizontal septum, wall bilaminate, the inner wall pigmented, the outer hyaline, germ pores 3, rarely 4, in each cell, sieve-like in appearance and apparently with multiple perforations, spores borne singly on pedicels, basidium external.

TYPE: Macruropyxis fraxini (Kom.) Azbu. (Puccinia fraxini Kom.).

This monotypic genus is similar to and perhaps not separable from Dipyxis on significant features. M. fraxini is microcyclic and has type 7 spermogonia. The spermogonia of Dipyxis have been too old to classify with certainty, but type 3 or perhaps 7 were reported originally. A possible difference of spermogonial types and the sieve-like appearance of the germ pores are all that separate Macruropyxis and Dipyxis.

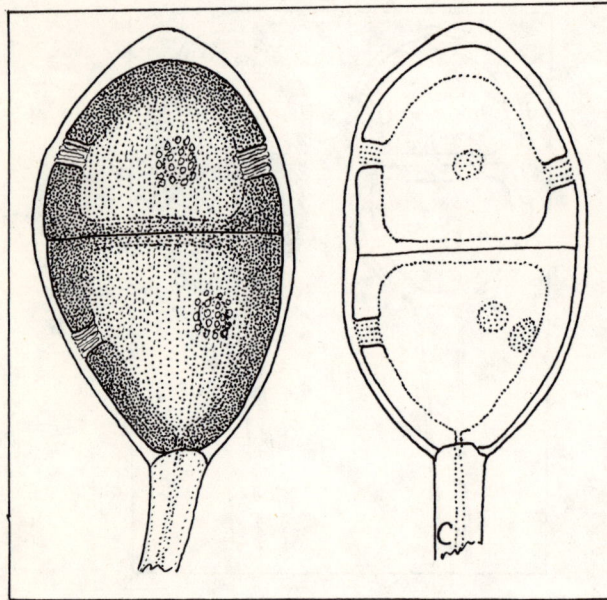

M. fraxini (Kom.) Azbu.; teliospores.

75. DIPYXIS Cummins & J.W. Baxter, Mycologia 59:368. 1967.

Spermogonia subcuticular, type 3 or 7? Aecia subepidermal in origin, erumpent, uredinoid; spores borne singly on pedicels, as the urediniospores. Uredinia subepidermal in origin, erumpent; spores borne singly on pedicels, echinulate, pores zonate. Telia subepidermal in origin, erumpent; spores 2-celled by horizontal septum, wall pigmented, germ pores 3, rarely 4, per cell, the spores borne singly, basidium external.

TYPE: Dipyxis mexicana Cumm. & J.W. Baxt.

The type occurs on Adenocaulon calderonii in Mexico. There is one other species, D. viegasii on Arrabidaea sp. in Brazil. Both hosts are Bignoniaceae and both rusts have echinulate teliospores and reniform urediniospores, i. e., with a concave side with germ pores near the hilum or strongly subequatorial.

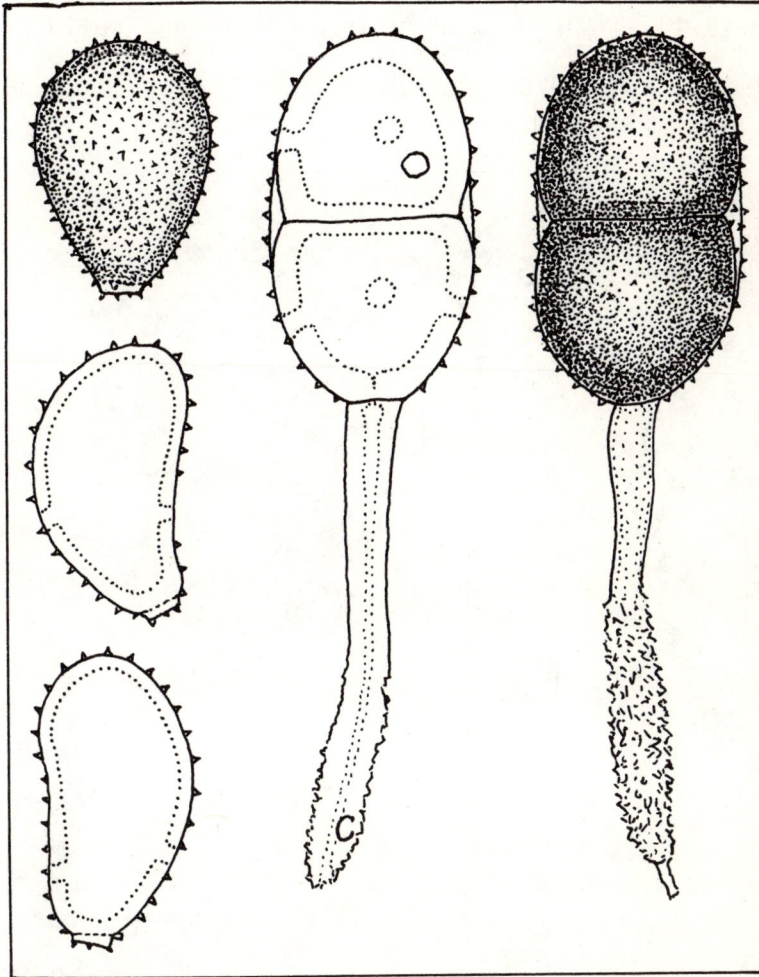

D. mexicana Cumm. & J.W. Baxt.; teliospores and urediniospores.

76. PHRAGMOPYXIS Dietel, Engler & Prantl Nat. Fam 1(1**):70. 1897

Spermogonia subcuticular, type 7. Aecia subepidermal in origin, erumpent, caeomatoid with catenulate, verrucose spores or uredinoid with echinulate spores borne singly. Uredinia subepidermal in origin, erumpent; spores borne singly on pedicels, echinulate, pores scattered, obscure. Telia subepidermal in origin, erumpent; spores borne singly on usually hygroscopic pedicels, mostly 3-celled by horizontal septa, wall conspicuously bilaminate, the outer layer pale to colorless and easily separable, germ pores 3 or 4 in each cell, pedicel often swollen basally and sometimes dissolving into the mounting medium, basidium external.

TYPE: Phragmopyxis deglubens (Berk. & Curt.) Diet. (Triphragmium deglubens Berk. & Curt.).

There are four species, all autoecious and all on legumes. P. acuminata is microcyclic. P. leonensis, from Sierra Leone, is the only extra-American species. It differs from others in having subcuticular uredinia and telia.

REFERENCE: Cummins, G.B. 1978. Rust fungi on legumes and composites in North America. Univ. Arizona Press. 424 pp.

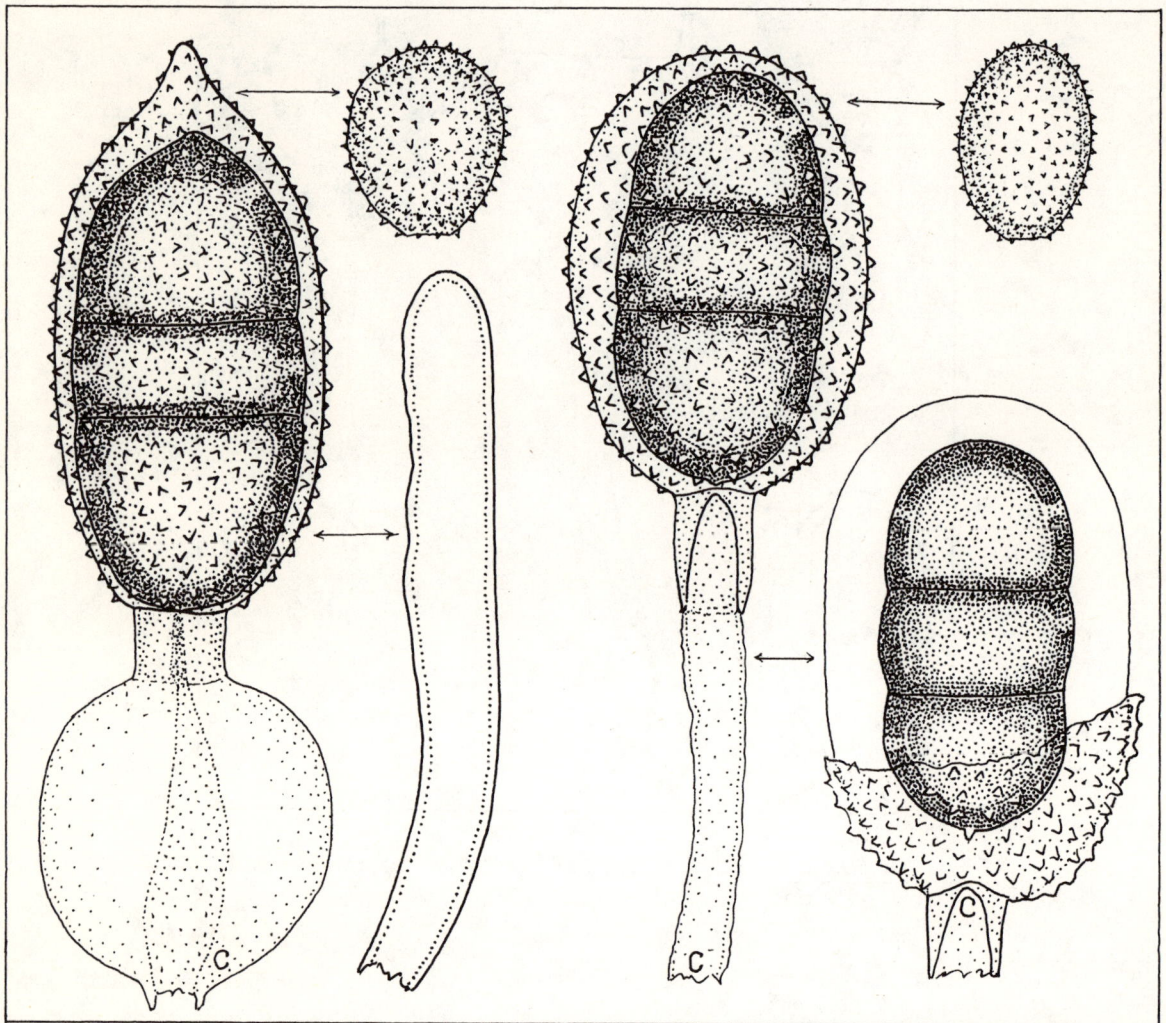

P. noelii J.W. Baxt.; one teliospore, one paraphysis, and one urediniospore (left). P. deglubens (Berk. & Curt.) Diet.; two teliospores and one urediniospore (right).

77. TRACHYSPORA Fuckel, Bot. Zeit. 19:250. 1861.

Spermogonia type 10 but usually not produced or not collected. Aecia subepidermal in origin, erumpent, without peridium; spores catenulate but without intercalary cells, appearing irregularly verrucose under moderate magnification but actually spinose (Henderson) with spines isolated or in groups on basal palques. uredinia not produced but some supposed urediniospores occasionally occur in the telia, echinulate, pores obscure. Telia subepidermal in origin, erumpent, produced both on systemic mycelium in the old aecia or localized; spores 1-celled borne singly on 1-septate pedicels, the upper cell short, wall of spore pigmented, germ pore obscure, basidium external.

TYPE: Trachyspora intrusa (Grev.) Arth. (T. alchemillae Fckl.).

Trachyspora is a genus of four species on Alchemilla of the Rosaceae. Henderson reports that the surface sculpture is similar to that of some species of Phragmidium, thus fortifying the opinion that the spores are aeciospores and that the relationship is with that genus. T. vestita (Diet.) Lindq. is a Hyphomycete, Chlamydomyces palmarum (Cke.) Mason (see Lindquist, Bol. Soc. Argent. Bot. 7:17. 1958.

REFERENCES: Gjaerum, H.B. and Cummins, G.B. 1982. Rust fungi (Uredinales) on East African Alchemilla. Mycotaxon 15:420-424. Henderson, D.M. 1973. Studies in the morphology of fungal spores, Trachyspora intrusa. Rept. Tottori Mycol. Inst. Japan. 10:163-168.

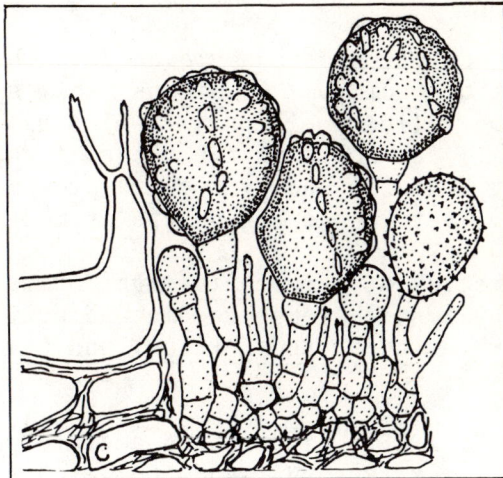

T. intrusa (Grev.) Arth.; telio-spores and one urediniospore.

113

78. ARTHURIOMYCES Cummins & Y. Hiratsuka, gen. nov.

     Spermogoniis intraepidermalibus, typus 6. Aeciis caeomatoideis, sporae
catenulatae, verrucosae. Urediniis nullis. Teliis subepidermalibus, erumpentibus,
sporae uniseptatae, pedicellatae, in quaque cellula poro germinationis unicus.
     Spermogonia intraepidermal, type 6. Aecia caeomatoid, without peridium or
paraphyses, indeterminate; spores catenulate, verrucose. Uredinia not produced.
Telia subepidermal in origin becoming erumpent; spores borne singly on pedicel, 2-
celled, wall pigmented, pore 1 in each cell.

     TYPE: Arthuriomyces peckianus (Howe in Peck) Cumm. & Y. Hirat., comb. nov.
(Puccinia peckiana Howe).

     This is the only species. The genus is provided as a replacement for
Gymnoconia, which Laundon determined to apply to the endocyclic form (see below).

63B. GYMNOCONIA Lagerheim, Troemso Mus. Aarsh. 16:142. 1894.

     Spermogonia as in Arthuriomyces or lacking. Aecia and uredinia lacking. Telia
as the aecia of Arthuriomyces (and to be distinguished with certainty only by germinating the spores), teliospores unicellular, catenulate, verrucose, basidium
external.

     TYPE: Gymnoconia nitens (Schw.) Kern & Thur.

     This microcyclic form has often been segregated in the genus Kunkelia Arth.
But Laundon has concluded that Aecidium "A. nitens Schw. applies to the perfect
state because it is based on telial aeciospores (which give rise to basidia (Art.
59) from N. Carolina (U.S.A.) far south of the range of the long cycle form."
Further, she concluded that Kunkelia and Gymnoconia are obligatory synonyms and
that if the long cycle form is to be kept separate it must have a new name (see
above).
     It is possible that species of Gymnoconia might arise by endocyclic abbrevia-
tion of the lifecycle of any rust that has caeomatoid aecia, e.g. Mikronegeria
or Melampsora. Hence, genera as this, Endophyllum, and Endocronartium are in a
sense pseudo-genera.

     REFERENCES: Dodge, B.O. 1925. Uninucleated aecidiospores in Caeoma nitens,
and associated phenomena. J. Agr. Res. 28:1045-1058. Kunkel, L.O. 1913. The
production of a promycelium by aeciospores of Caeoma nitens Burrill. Bull. Torrey
Bot. Club 40:361-366. Laundon, G.F. 1975. Taxonomy and nomenclature notes on
Uredinales. Mycotaxon 3:133-161.

                          Illustrations next page.

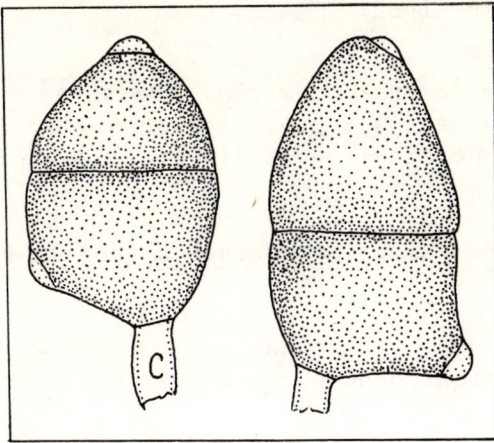

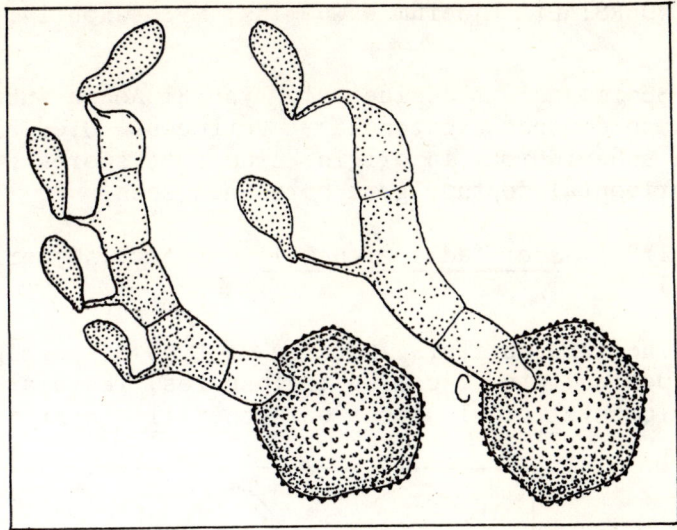

A. peckianus (Howe) Cumm. & Y. Hirat.; teliospores (left). G. nitens (Schw.)
Kern & Thur.; teliospores (right).

79.   JOERSTADIA Gjaerum & Cummins, Mycotaxon 14:420. 1982.

Spermogonia subepidermal, type 8. Aecia subepidermal in origin, erumpent, caeomatoid; spores catenulate, verrucose. Uredinia unknown, probably not produced. Telia subepidermal in origin, erumpent; spores borne singly on pedicels, 2-celled by horizontal septum, germ pores not seen.

TYPE: Joerstadia alchemillae (Bacc.) Gjaer. & Cumm. (Gymnoconia alchemillae Bacc.).

The genus differs, because of type 8 spermogonia, from other genera that have pedicellate, 2-celled teliospores. There are four species, two demicyclic and two microcyclic; all occur on Alchemilla in Africa.

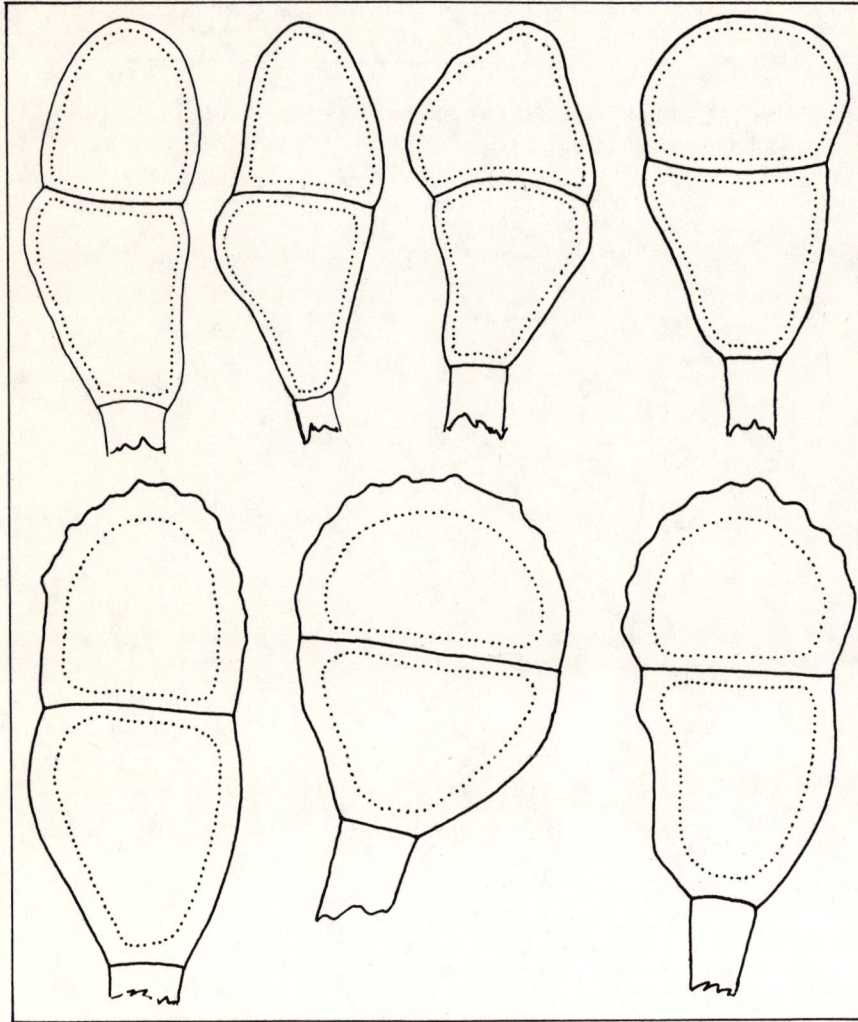

J. alchemillae (Bacc.) Gjaer. & Cumm.; teliospores (above). J. keniensis Gjaer. & Cumm.; teliospores (below). Both adapted from Gjaerum & Cummins, Mycotaxon 1982.

80. HAMASPORA Koernicke, Hedwigia 16:22. 1877.

Spermogonia sunbepidermal, type 8 or 10. Aecia subepidermal in origin, erumpent, uredinoid; spores borne singly on pedicels, echinulate. Uredinia subepidermal in origin, erumpent, paraphysate; spores borne singly on pedicels, echinulate, pores obscure. Telia subepidermal in origin, erumpent in pale, felt-like or yarn-like masses; spores 2- several-celled by horizontal septa, borne singly on stout pedicels, spore wall pale and more or less uniformly thin except the apex, germ pore probably 1 per cell but obscure, germination occurs without dormancy, basidium external.

TYPE: Hamaspora longissima (Thuem.) Koern. (Phragmidium longissimum Koern.).

The 15 species are presumed to be autoecious; all occur on Rosaceae; and all occur in warm areas from Africa to Australia, Japan, and the Philippines. The species have long nematode-like teliospores. The genus presumably is related to Phragmidium but has different spermogonia and pale teliospores presumably with a single germ pore per cell.

REFERENCE: Monoson, H.L. 1969. The species of Hamaspora. Mycopathol. Mycol. Appl. 37:263-272.

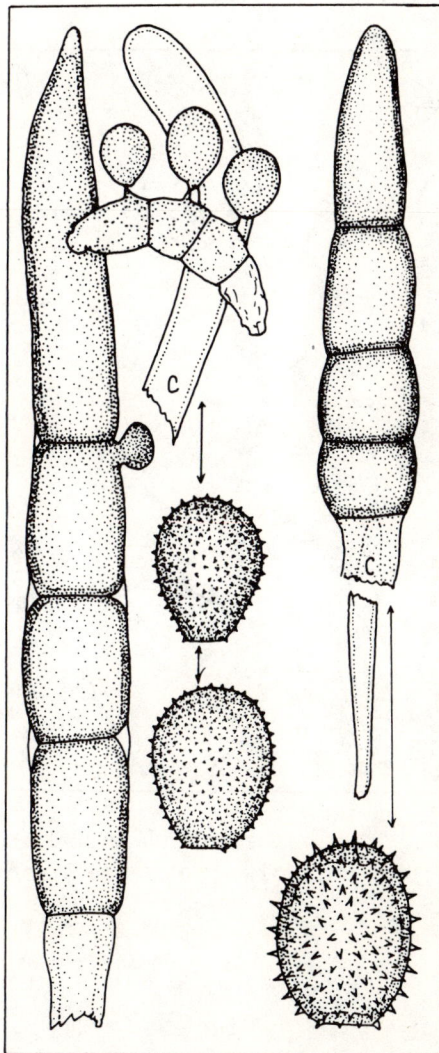

H. longissima (Theum.) Koern.
(left). H. hashiokae Hirat. f.
(right).

81. KUEHNEOLA Magnus, Bot. Zentralbl. 74:169. 1898.

Spermogonia subcuticular, type 11. Aecia subepidermal in origin, erumpent, uredinoid; spores borne singly on pedicels. Uredinia subepidermal in origin, erumpent; spores borne singly on pedicels, echinulate or otherwise marked, pores equatorial or obscure. Telia subepidermal in origin, erumpent; spores borne singly on short pedicels, 2- to several-celled by horizontal septa, wall pale or colorless, smooth, germ pores 1 in each cell, germination occurs without dormancy, basidium external.

TYPE: Kuehneola uredinis (Link) Arth. (Chrysomyxa albida Kuehn.).

The genus has only a few species; most are on Rubus (Rosaceae); all are autoecious. K. uredinis may cause defoliation of commercial plantings and also may infect stems, causing "cane blight."

REFERENCES: Arthur, J.C. 1912. Uredinales. N. Amer. Flora 7:182-187 (incl. Spirechina in part). Jackson, H.S. 1931. The rusts of South America based on the Holway collections -III. Mycologia 23:96-116 (see pp. 105-106). For surface sculpture of urediniospores see SEM micrographs in López, A., Carrion, G., Galvan, M., and Martínez, D. in Bol. Soc. Mex. Mic. 14:81-100. 1980.

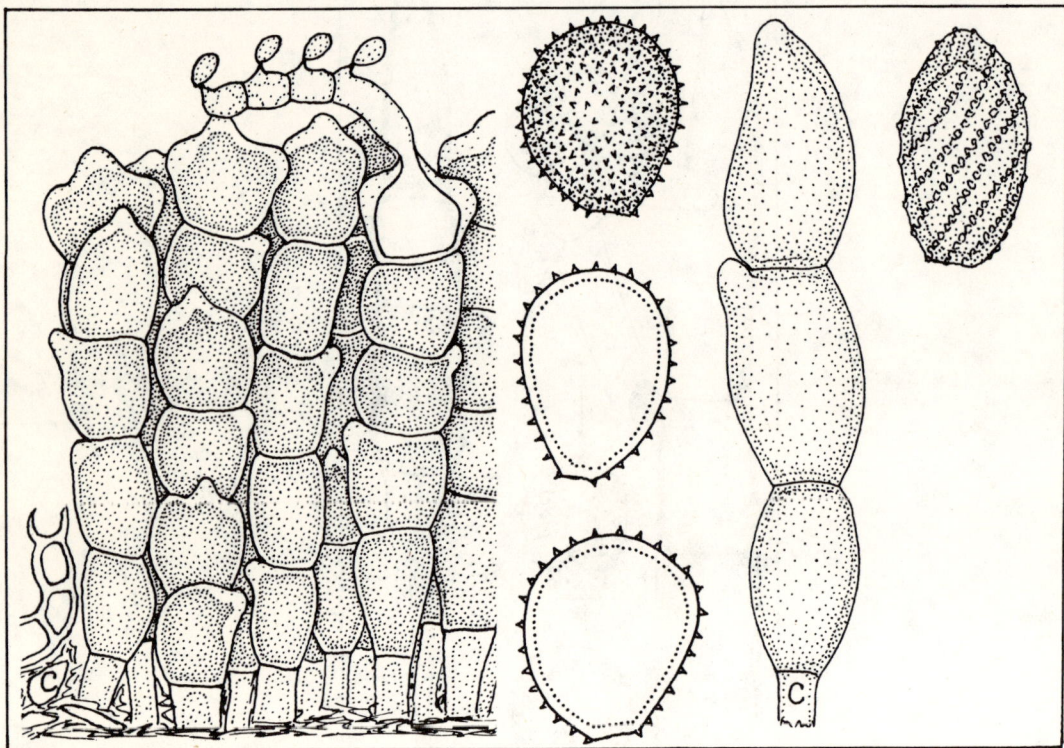

K. uredinis (Link) Arth.; teliospores and three urediniospores (left).
K. loesneriana (Arth.) Jack. & Holw.; one teliospore and one urediniospore (right).

82. NEWINIA Thaung, Mycologia 65:702. 1973.

Spermogonia subcuticular, type 7. Aecia subepidermal in origin, erumpent, uredinoid; spores borne singly on pedicels, essentially as the urediniospores. Uredinia subepidermal in origin, erumpent; spores borne singly on pedicels, echinulate, pores equatorial. Telia subepidermal in origin, erumpent; spores borne singly on pedicels, 3- to many-celled by horizontal septa, wall pigmented, smooth, germ pore 1, apical in terminal cell, 1 less often 2 next the septum in other cells, basidium external.

TYPE: Newinia heterophragmae Thaung.

The genus and type species were described from Burma on Heterophragma sulphureum but the same or a very closely related fungus occurs in Nigeria on Kigelia africana. Both plants belong in the Bignoniaceae. The teliospores are somewhat as those of Kuehneola, Frommeëlla, and Xenodochus but these have type 10 or 11 spermogonia. The urediniospores of Newinia are much more elaborate than those of the other genera.

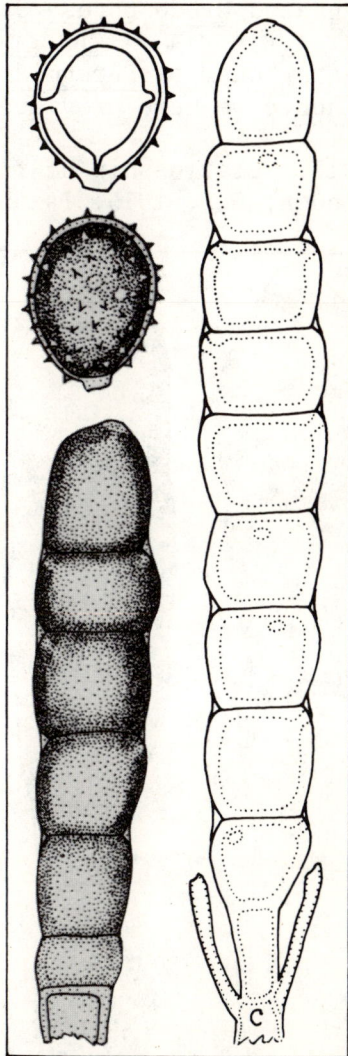

N. heterophragmae Thaung; urediniospores and teliospores.

83. FROMMEËLLA Cummins & Y. Hiratsuka, gen. nov.

Spermogoniis intraepidermalibus, typus 10. Aeciis Uraecium. Urediniis aeciis similes. Teliis subepidermalibus deinde erumpentibus, sporae plerumque 3 vel 4 loculatae, brunneae, pedicellatae, poro germinationis in quisque cellula unicus.

Spermogonia intraepidermal, type 10. Aecia subepidermal in origin, erumpent, uredinoid; spores borne singly, echinulate, pores equatorial, obscure. Uredinia subepidermal in origin, erumpent; spores similar to aeciospores. Telia subepidermal in origin, erumpent; spores borne singly on short pedicels, 3- to several-celled by horizontal septa, wall pigmented, smooth, germ pore 1 per cell, germinating without dormancy, basidium external.

TYPE: Frommeëlla tormentillae (Fckl.) Cumm. & Y. Hirat. comb. nov. (Phragmidium tormentillae Fckl.).

The species of Frommeëlla are autoecious and restricted to Potentilla and close relatives of the Rosaceae. The genus differs from Phragmidium because of the single germ pore per teliospore cell, but doubtless is closely related. When Arthur described Frommea he designated Uredo obtusa Strauss as the type. Laundon has pointed out that U. obtusa is a synonym of Phragmidium potentillae and by designated typification Frommea is a synonym of Phragmidium. But the fungus described by Arthur is not a Phragmidium, hence we here provide a new generic name.

REFERENCES: Arthur, J.C. 1917. Relationship of the genus Kuehneola. Bull. Torrey Bot. Club 44:501-511. Laundon, G.F. 1975. Taxonomy and nomenclature notes on Uredinales. Mycotaxon 3:133-161.

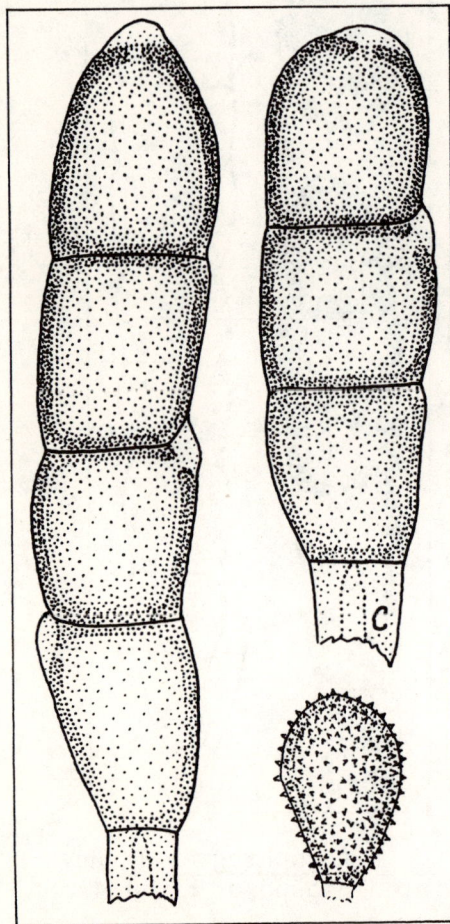

F. tormentillae (Fckl.) Cumm.
and Y. Hirat.; teliospores and
one urediniospore.

84.   XENODOCHUS Schlechtendahl, Linnaea 1:237. 1826.

Spermogonia intraepidermal, type 10. Aecia subepidermal in origin, erumpent, caeomatoid; spores catenulate, with papillate verrucae. Uredinia and spores similar to aecia and aeciospores except not accompanied by spermogonia. Telia subepidermal in origin, erumpent; spores borne singly on short pedicels, 2- to many-celled by horizontal septa, wall smooth, pigmented, pore 1 (rarely 2) in apical cell, others with 2 pores, basidium external.

TYPE: Xenodochus carbonarius Schl.

Two species are known, X. carbonarius and X. minor. Both are circumboreal on species of Sanguisorba of the Rosaceae. The type is autoecious and macro-cyclic; X. minor is microcyclic. The genus differs from Frommeëlla because of two germ pores in all but the apical teliospore cell. Sato and Sato have completed the life cycle of X. carbonarius by inoculations.

REFERENCE: Sato, T. and Sato, S. 1980. The caeomoid uredinium of Xenodochus carbonarius (Uredinales). Trans. Mycol. Soc. Japan 21:411-416.

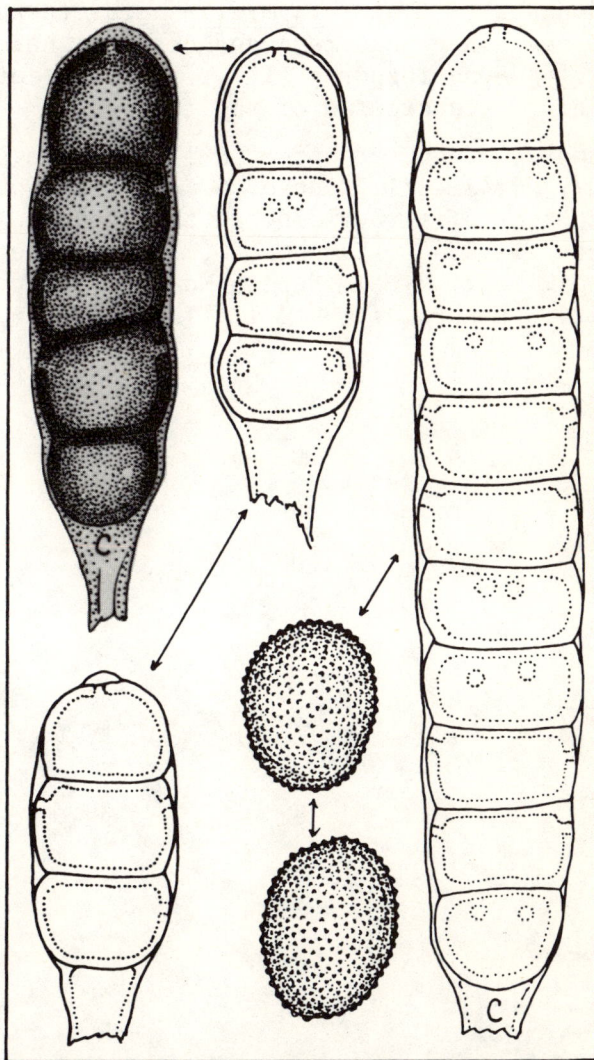

X. minor Arth.; three teliospores (left). X. carbonarius Schl.; teliospore and two urediniospores (right).

85.   PHRAGMIDIUM Link, Mag. Ges. Naturf. Freunde Berlin 7:30. 1816.

Spermogonia subcuticular, type 11 or intraepidermal, type 10. Aecia subepidermal in origin, erumpent, caeomatoid with catenulate spores or less often uredinoid with spores borne singly; spores verrucose or echinulate, pores scattered. Uredinia subepidermal in origin, erumpent, with peripheral paraphyses; spores borne singly on pedicels, mostly echinulate, pores scattered, obscure. Telia subepidermal in origin, erumpent; spores borne singly on often hygroscopic pedicels, 1- to several-celled by horizontal septa, wall pigmented, smooth or more often verrucose, often obviously bilaminate, germ pores 2 or usually 3 in each cell, basidium external.

TYPE: Phragmidium mucronatum (Pers.) Schl. (Puccinia mucronata Pers.).

Some 60 to 65 species have been recognized but not all are distinctive. There is no world monograph of recent date. All species are autoecious and most are macrocyclic but there are demicyclic and microcyclic species. All species occur on the Rosaceae with Rosa, Rubus, and Potentilla the common hosts. The genus predominantly inhabits the Northern Hemisphere.

There are about ten species that have uredinoid aecia and terete, basally septate pedicels. These sometimes are separated in Phragmotelium, as by Thirumalachar and Mundkur. Laundon states that Phragmidium needs to be conserved against Aregma but this has not been done. Henderson and Prentice have described the formation of the spines of the aeciospores of P. fragariae, of the urediniospores of P. tuberculatum, and of the tuberculae of the latter.

REFERENCES: Henderson, D.M. and Prentice, H.T. 1973. Development of spores of Phragmidium. Nova Hedw. 24:431-441. Kaneko, S. and Nishigaki, H. 1980. A taxonomic revision of the species of Phragmidium in the Japanese Archipelago. Rept. Tottori Mycol. Inst. Japan. 18:53-88. Laundon, G.F. 1965. The generic names of Uredinales. Commonw. Mycol. Inst. Mycol. Papers No. 99. 24 pp. Thirumalachar, M.J. and Mundkur, B.B. 1949. Genera of rusts II. Indian Phytopathol. 2:193-244 (see p. 243).

Illustrations next page.

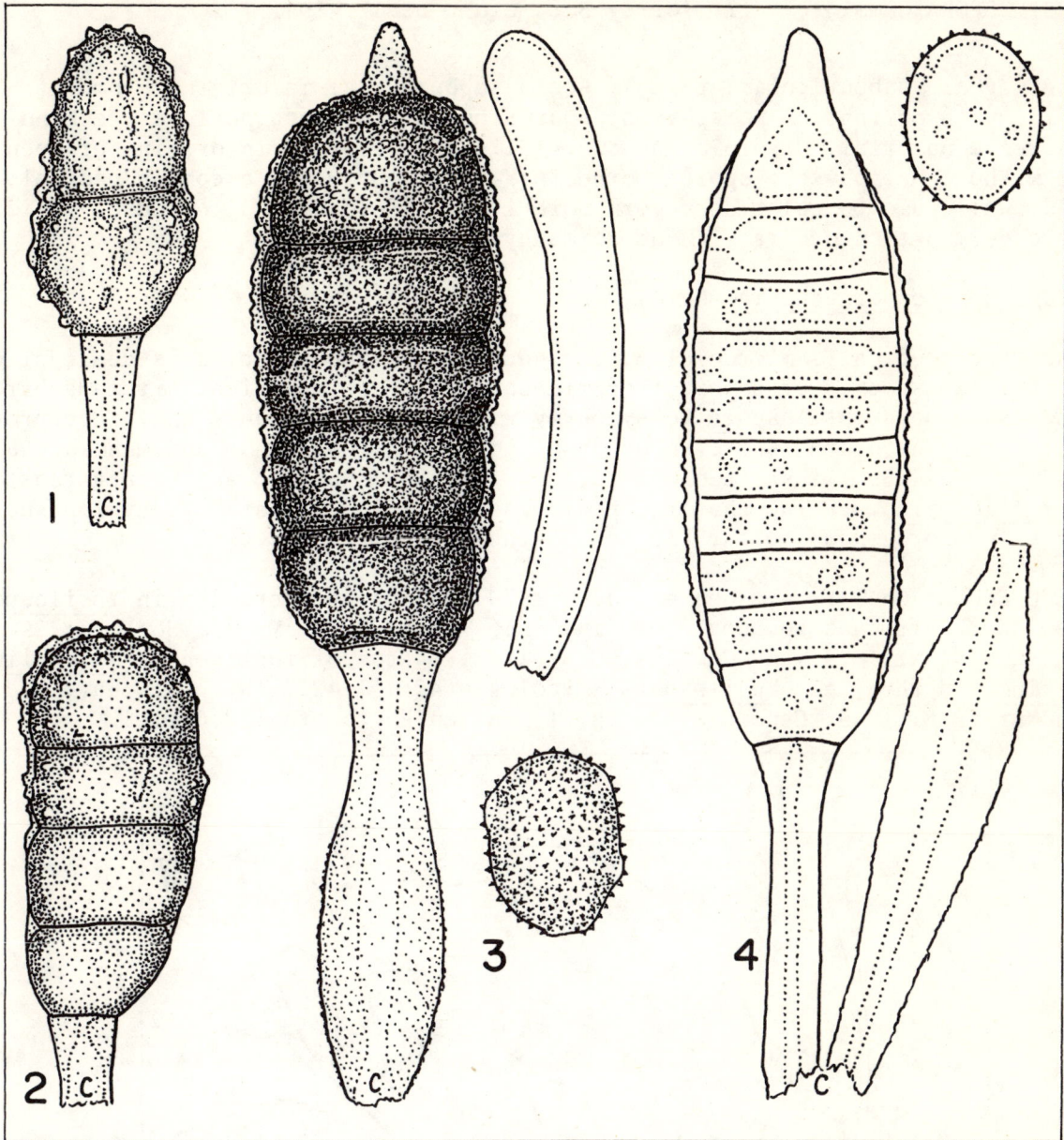

Phragmidium species. 1. _P. biloculare_ Diet. & Holw.; a teliospore. 2. _P. boreale_ Tranz.; a teliospore. 3. _peckianum_ Arth.; a teliospore, a paraphysis, and one urediniospore. 4. _P. fusiforme_ Schroet.; a teliospore and one urediniospore.

86. YPSILOSPORA Cummins, Bull. Torrey Bot. Club. 68:47. 1941.

Spermogonia subcuticular, type 7. Aecia subepidermal in origin, erumpent, uredinoid, with peripheral paraphyses; spores borne singly on pedicles, echinulate, pores equatorial. Uredinia unknown. Telia subepidermal in origin, erumpent, with or without paraphyses; spores borne in Y-shaped pairs on a common pedicel, 1-celled, wall pale to colorless, germ pore 1 if differentiated, obscure, basidium external, germination occurs without dormancy.

TYPE: Ypsilospora baphiae Cumm.

The type species is microcyclic; the second described species is demicyclic. Both parasitize species of Baphia (Leguminosae) in western Africa.Thirumalachar and Cummins reduced Ypsilospora to synonymy with Chaconia because they interpreted the pedicels as elongated sporogenous basal cells. Thirumalachar and Mundkur, and Cummins, in the 1959 Illustrated Genera, did likewise. But Ono and Hennen reestablished Ypsilospora, stating that the pedicels are pedicels that "...develop successively on sporogenous cells." We accept their interpretation.

REFERENCES: Ono, Y. and Hennen, J.F. 1979. Teliospore ontogeny in Ypsilospora baphiae and Y. africana sp. nov. (Uredinales). Trans. Brit. Mycol. Soc. 73:229-233. Thirumalachar, M.J. and Cummins, G.B. 1948. Status of rust genera Allotelium, Leucotelium, Edythea, and Ypsilospora. Mycologia 40:417-422. Thirumalachar, M.J. and Mundkur, B.B. 1949. Genera of rusts. I. Indian Phytopathol 2:65-101.

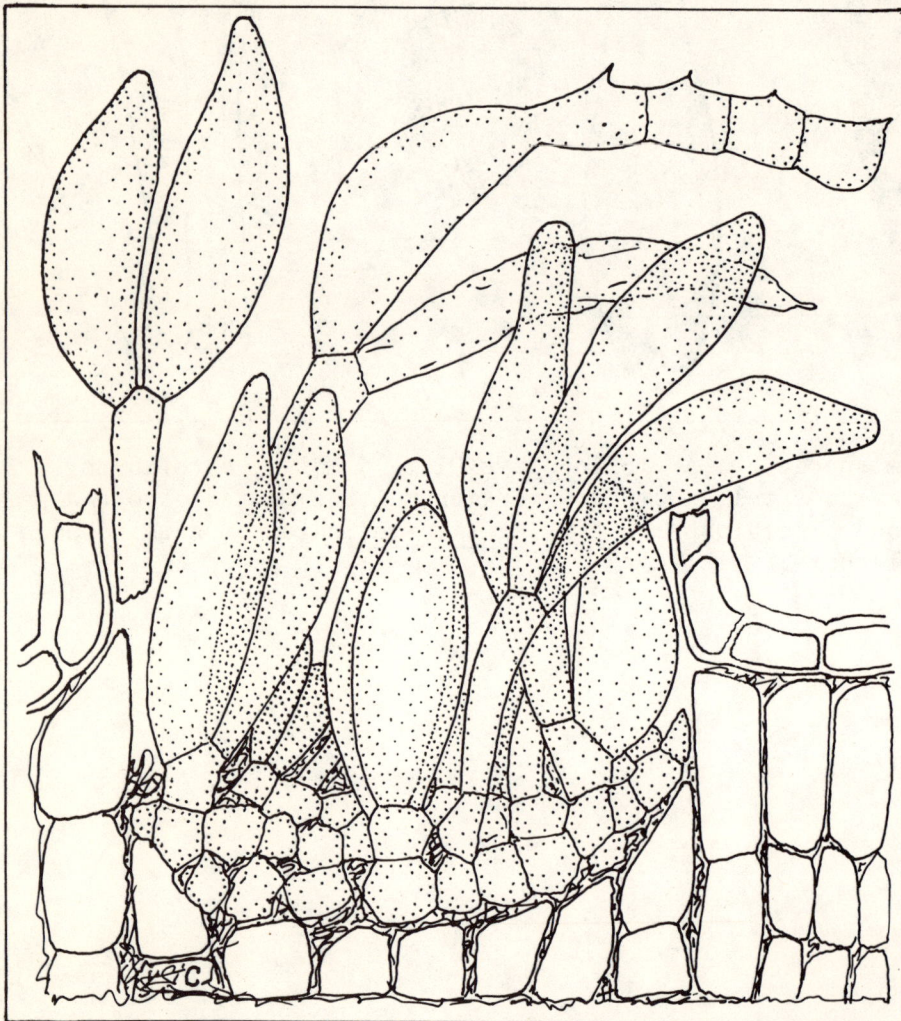

Y. baphiae Cumm.; teliospores.

87. APRA Hennen & Freire, Mycologia 71:1054. 1979.

Spermogonia splitting the epidermis, type 7? Aecia subepidermal, deep-seated, becoming erumpent, with peridium, aecidioid; spores catenulate. Uredinia unknown. Telia subepidermal in origin, with a marginal layer of compacted hyphae and collapsed palisade cells, becoming erumpent; spores 1-celled, borne in laterally free pairs on a common pedicel which has 2 apical cells, 1 for each spore, germ pore not seen but basidium doubtless external.

TYPE: Apra bispora Hennen & Freire.

Only the type species is known. It occurs on Mimosa micrantha in Brazil. Apra differs from Dicheirinia and Diorchidiella, each of which has one apical cell for each spore, because the spores are free from one another. Diabole, also on Mimosa, has two free spores on a single apical cell and with one or more pairs on a common pedicel.

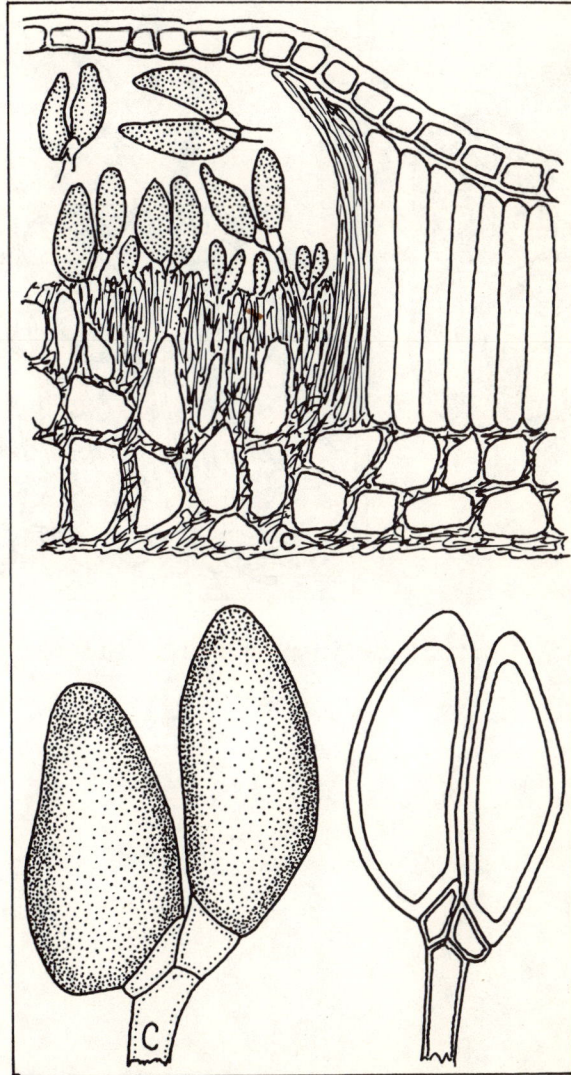

A. bispora Hennen & Freire; telium and teliospores.

88.   DIABOLE Arthur, Bull. Torrey Bot. Club 49:194. 1922.

Spermogonia subcuticular, type 7. Aecia and uredinia unknown. Telia subcuticular in origin, erumpent; spores borne singly on apical cells of a pedicel
either as a single pair or typically as 2 or 3 pairs, thus 4 or 6 pairs of 1-cel-
led spores, wall pigmented, verrucose, germ pores uncertain, perhaps in a basal
pale area of each spore, germination unknown but the basidium undoubtedly external.

TYPE: Diabole cubensis (Arth.) Arth. (Uromycladium cubense Arth.).

Only the type species is known. It occurs in Cuba, Central America, Mexico,
and Brazil on Mimosa (Leguminosae). Diabole is one of a few genera that have subcuticular telia. The genus presumably is related to Ravenelia or Dicheirinia, or
perhaps Uromycladium where Arthur first placed it.

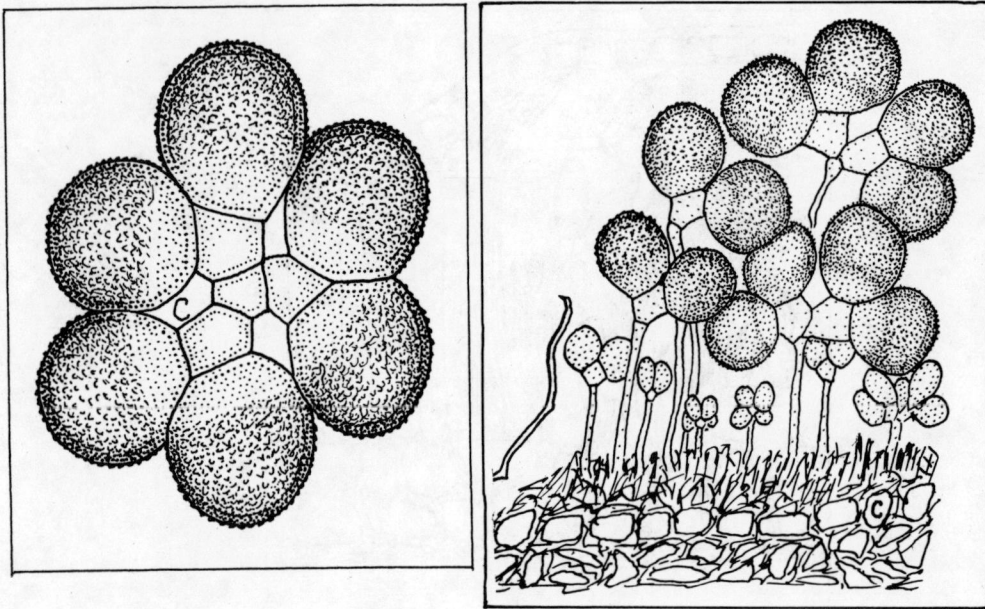

D. cubensis (Arth.) Arth.; teliospores and telium.

89. DIORCHIDIELLA Lindquist, Darwiniana 11:416. 1957.

Spermogonia, aecia, and uredinia unknown. Telia subepidermal in origin, erumpent; spores 2-celled by vertical septum, borne singly on pedicels, the pedicel having 2 apical cells, 1 for each teliospore cell, wall pigmented, germ pores 2 in each cell; basidium external.

TYPE: Diorchidiella australis (Speg.) Lindq. (Diorchidium australe Speg.).

Only the type species is known. It occurs in South America on Mimosa. The genus is separable from Dicheirinia because of the two germ pores per cell.

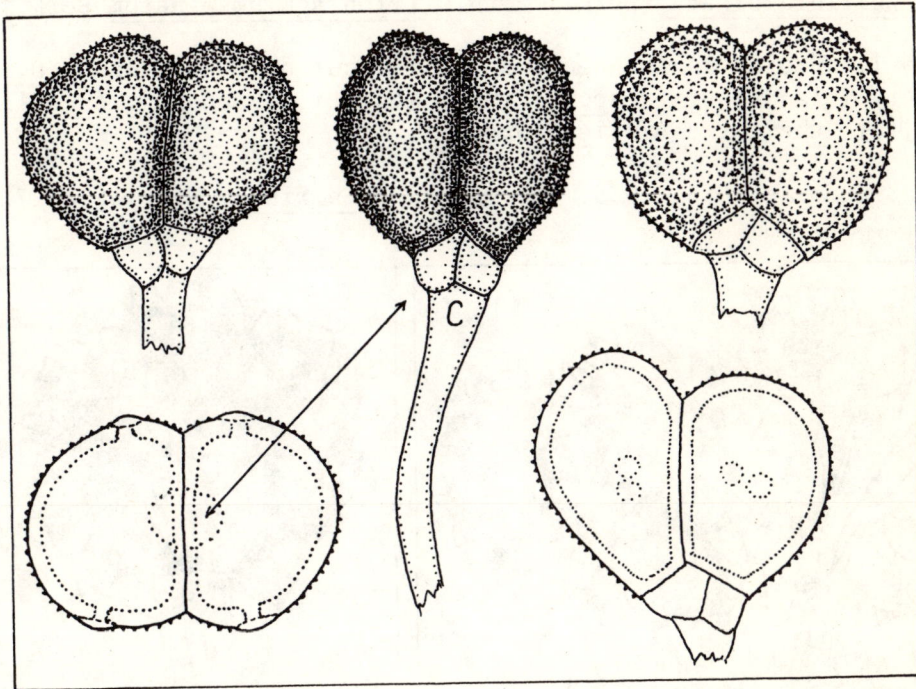

D. australis (Speg.) Lindq.; teliospores.

90. DICHEIRINIA Arthur, N. Amer. Flora 7:147. 1907.

Spermogonia subcuticular, type 7. Aecia subepidermal in origin, erumpent, uredinoid; spores borne singly on pedicels. Uredinia subepidermal in origin, erumpent, similar to the aecia but without spermogonia, mostly with peripheral paraphyses which often are branched and "fancy"; spores borne singly on pedicels, as the aeciospores, pores equatorial or basal. Telia subepidermal in origin, erumpent; spores 2-4-celled by vertical septa, wall pigmented, mostly ornamented with block-like warts, the pedicel with 1 apical cell for each cell of the spore, germ pore 1 in each cell, basidium external.

TYPE: Dicheirinia binata (Berk.) Arth. (Triphragmium binatum Berk.).

There are 11 species in the genus and all occur on Leguminosae. D. canariensis Urr., Canary Islands, D. trispora Cumm., Mauritius, and D. viennotii Hugu., New Caledonia, are the extra-American species.

REFERENCE: Cummins, G.B. 1935. The genus Dicheirinia. Mycologia 27:151-159.

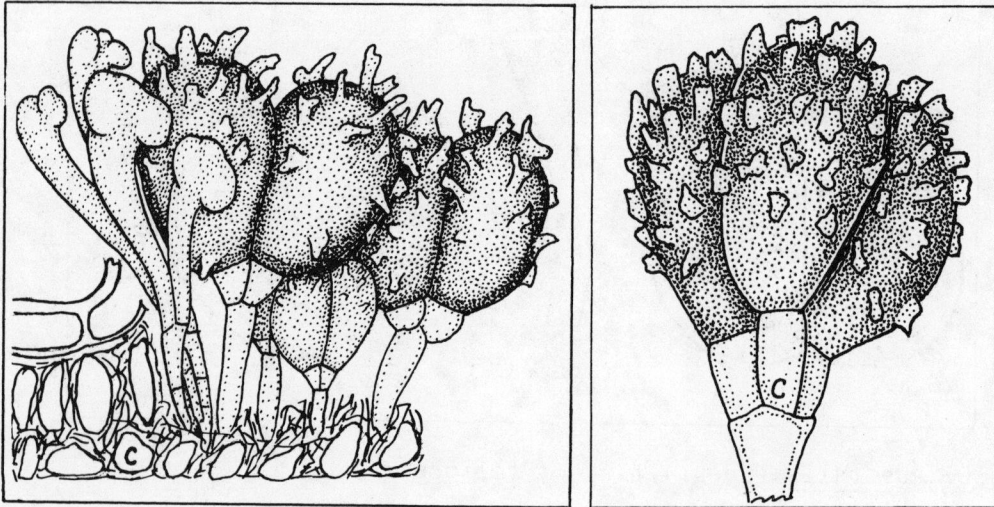

D. binata (Berk.) Arth.;paraphyses and teliospores.(left).
D. manaosensis (P. Henn.) Cumm.; a teliospore (right).

91. ANTHOMYCES Dietel, Hedwigia 38:253. 1899.

Spermogonia and aecia unknown. Uredinia subepidermal in origin, erumpent, with peripheral paraphyses; spores borne singly on pedicels, echinulate, pores obscure. Telia similar to the uredinia; spores 1-celled, firmly united laterally to form a radial cluster of 3- to several spores, wall pigmented, germ pore 1 in each spore, apical, pedicels with apical cells of the same number as the spores (or at least as the peripheral ones), basally simple, germination occurs without dormancy, basidium external.

TYPE: Anthomyces brasiliensis Diet.

The single species occurs in Brazil on an unidentified legume. The genus differs from Anthomycetella because the latter has multihyphal pedicels. The simple pedicel with apical cells is as that of Dicheirinia, although the smooth spores are different. If Anthomyces proves to have type 7 spermogonia and uredinoid aecia there would be little to separate the genera.

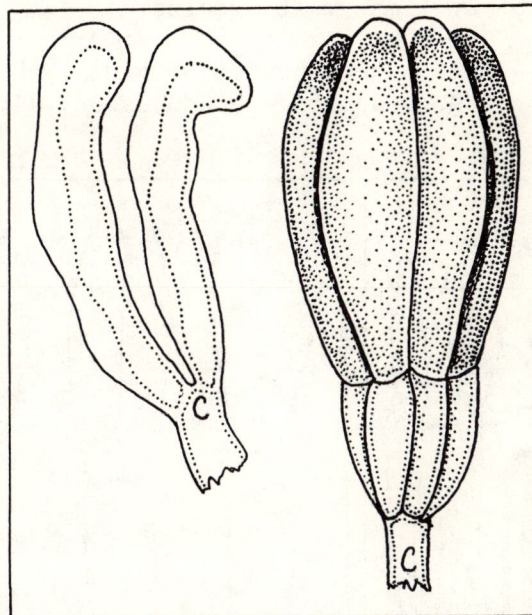

A. brasiliensis Diet.; paraphyses and teliospores.

92. SPHENOSPORA Dietel, Ber. Dtsch. Bot. Ges. 10:63. 1892.

Spermogonia and aecia unknown. Uredinia subepidermal in origin, becoming erumpent but often slowly; spores borne singly on pedicels, echinulate, pores equatorial or obscure. Telia subepidermal in origin, erumpent as waxy-oily cushions when moist, hard when dry; spores borne singly on pedicels, 2-celled by vertical septum, mostly narrowly ellipsoid or conical, wall pale to colorless, smooth, germ pore 1 in each cell if differentiated, apical, germination occurs without dormancy, basidium external.

TYPE: Sphenospora pallida (Wint.) Diet. (Diorchidium pallidum Wint.).

Nine of the ten species occur in tropical America and most are on monocots. S. xylopiae, on Annonaceae in Gabon, is the extra-American species. S. copaiferae, on Copaifera of the Leguminosae, we considered to belong to Diorchidium. Three species occur on orchids and often are intercepted by the United States Plant Quarantine Service in air cargo to the U. S.

An abundance of orange colored oil and an apparently gelatinous matrix is characteristic of the telia of at least most species.

REFERENCE: Linder, D.H. 1944. A new rust of orchids. Mycologia 36:464-468. Yen, J.-M. and Sulmont, P.H. 1969. Les Urédinées du Gabon. II. Bull. Soc. Mycol. France 85:351-353.

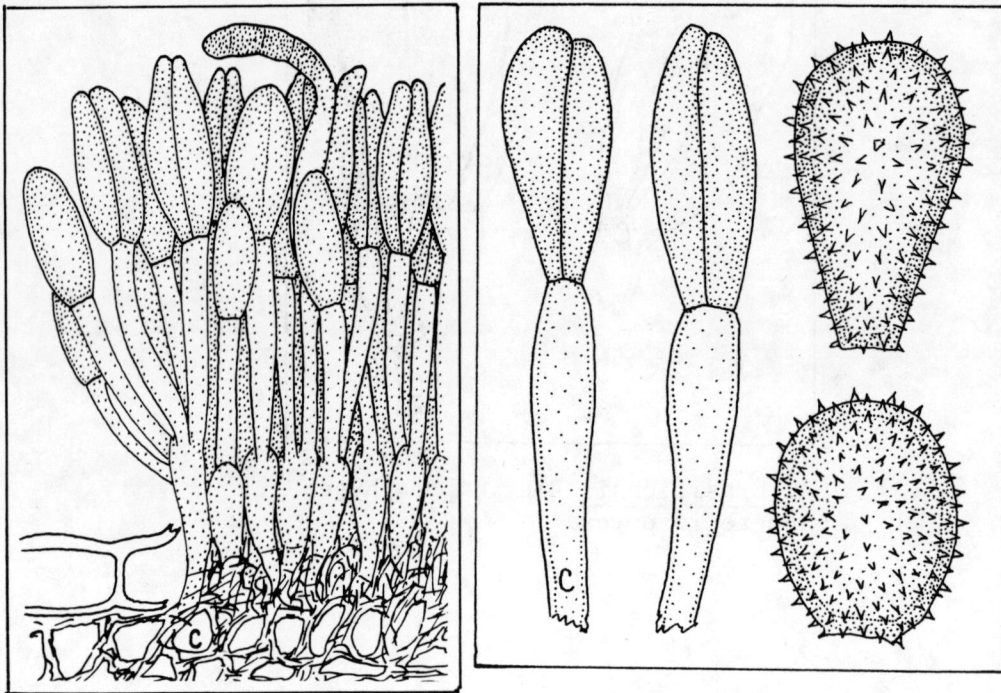

S. smilacina Syd.; sorus with teliospores (left). S. saphena Cumm.; teliospores and urediniospores (right).

93. DIORCHIDIUM Kalchbrenner, Grevillea 11:26. 1882.

Spermogonia subcuticular, type 7. Aecia subepidermal in origin, erumpent, uredinoid; spores borne singly on pedicels, echinulate, pores equatorial. Uredinia subepidermal in origin, erumpent; spores as the aeciospores. Telia subepidermal in origin, erumpent; spores 2- or less commonly 4-celled by vertical septa, borne singly on pedicels, wall pigmented, germ pore 1 in each cell, apical, basidium external.

TYPE: Diorchidium woodii Kalch. & Cooke.

This is a genus of few species, the number uncertain because species of Puccinia with more or less vertical septa have been described as Diorchidium. The spermogonia of Puccinia are type 4 and the teliospores rarely have vertical septa. Cummins described D. quadrifidum and D. tetrasporum, both with 4-celled teliospores, thus broadening the concept of the genus. We believe that D. copaifera (Syd.) comb. nov (Sphenospora copaiferae Syd.) belongs here. It is similar to D. quadrifidum and D. tetrasporum but with two cells. Diphragmium Boedijn, based on Diorchidium koordersii Wurth, differs because the teliospore pore is apical. We do not consider this to be important.

REFERENCES: Boedijn, K.B. 1959. The Uredinales of Indonesia. Nova Hedw. 1: 463-496. Cummins, G.B. 1960. Descriptions of tropical rusts-IX. Torrey Bot. Club Bull. 87:31-45.

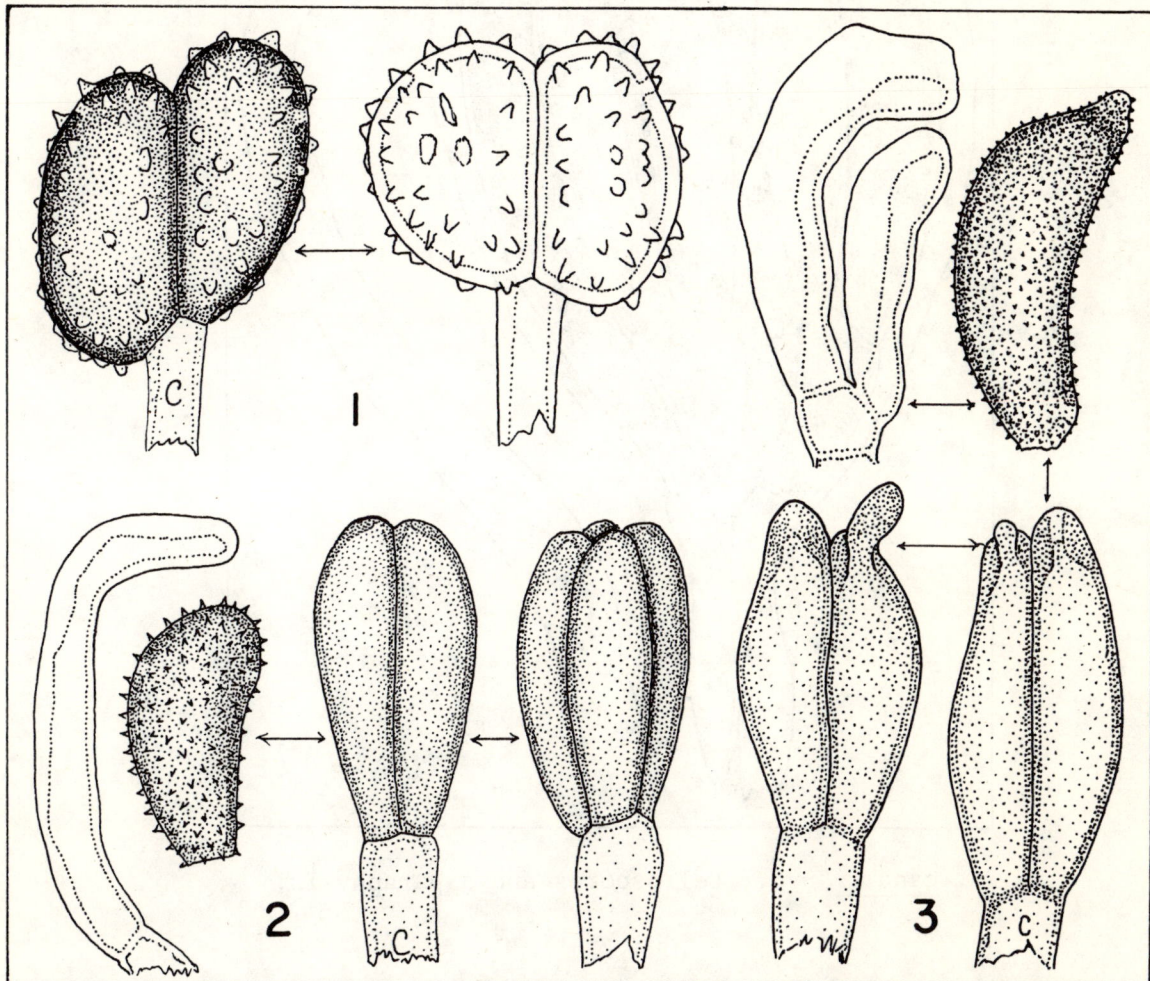

1. D. woodii Kalch. & Cke.; teliospores. 2. D. tetraspora Cumm.; a paraphysis, a urediniospore, and teliospores. 3. D. copaiferae (Syd.) Cumm. & Y. Hirat.; teliospores, paraphyses, and one urediniospore.

94. ANTHOMYCETELLA H. & P. Sydow, Ann. Mycol. 14:353. 1916.

Spermogonia and aecia unknown. Uredinia subepidermal in origin, erumpent, with peripheral paraphyses; spores borne singly on pedicels, echinulate, pores obscure. Telia as the uredinia; spores 1-celled, firmly united to form a radial cluster 2 to several spores wide, pigmented, overlaid by a pale layer and subtended by short spore-like basal cells, germ pore 1 in each spore, apical, pedicel compound corresponding to the number of basal cells, germination occurs without dormancy, basidium external.

TYPE: Anthomycetella canarii Syd.

The known species occurs on Canarium (Burseraceae) in the Philippines. Thirumalachar demonstrated that the basal layer of cells are not spores but sporogenous cells, each capable of producing one or more spores.

REFERENCE: Thirumalachar, M.J. 1947. Brief notes on the genera Stereostratum Magn. and Anthomycetella Syd. Mycologia 39:334-340.

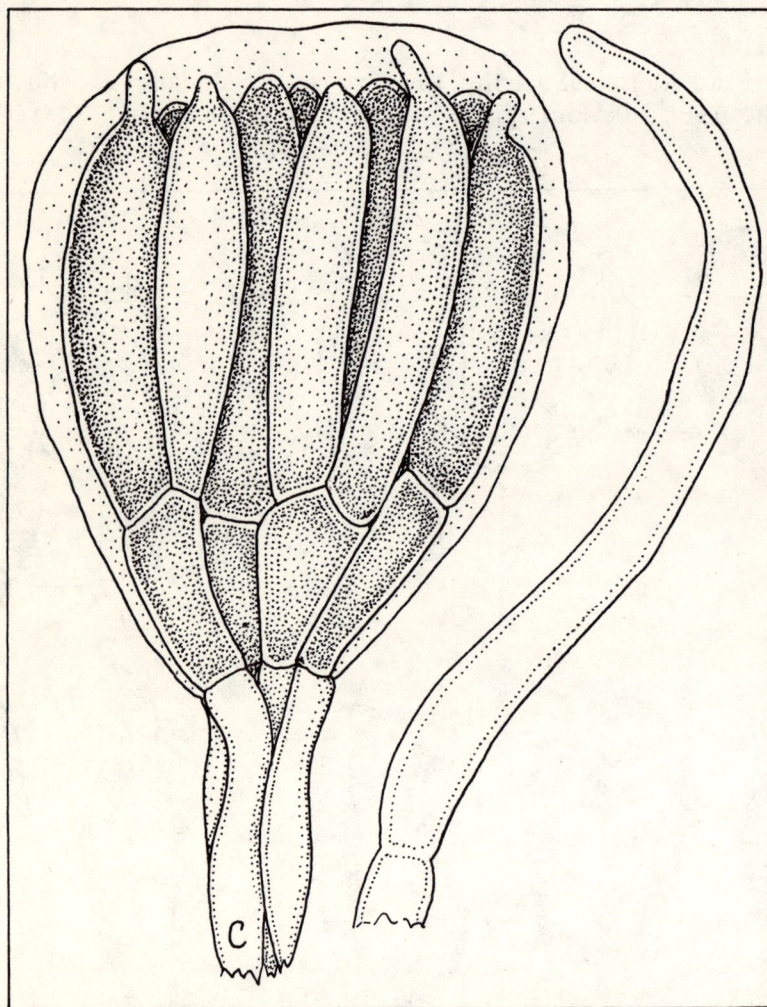

A. canarii Syd.; teliospores and a paraphysis.

95. NOTHORAVENELIA Dietel, Ann. Mycol. 8:310. 1910.

Spermogonia and aecia unknown. Uredinia subepidermal in origin, erumpent, echinulate, pores obscure. Telia subepidermal in origin, erumpent, with basally united, peripheral paraphyses; spores laterally and terminally strongly adherent forming discoid heads, the heads 1-3 spores thick, each chain of spores subtended by a cyst-like cell, the cystoid cells remain attached to the spore heads but separate from the sorus, successive heads may form and force the older ones from the sorus, wall of spores pigmented, cystoid cells hyaline, germ pores not seen, germination unknown but basidium undoubtedly external.

TYPE: Nothoravenelia japonica Diet.

There are two species, the type and N. commiphorae Cumm.; both occur on Euphorbiaceae. The genus is superficially similar to Ravenelia and Kernkampella but the relationship probably is with the phakopsoroid genera.

REFERENCES: Cummins, G.B. 1952. Uredinales from various regions. Bull. Torrey Bot. Club 79:212-234. Thirumalachar, M.J. and Mundkur, B.B. 1949. Genera of rusts I. Indian Phytopathol. 2:65-101.

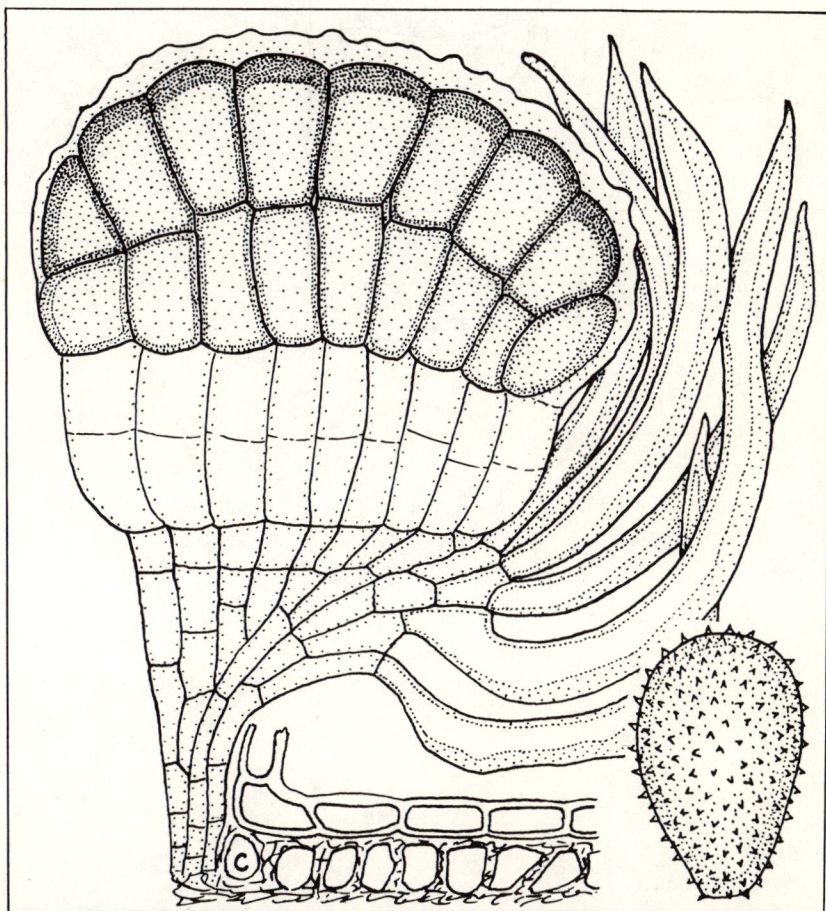

N. commiphorae Cumm.; a teliospore head with paraphyses and one urediniospore.

96. CYSTOMYCES H. Sydow, Ann. Mycol. 24:290. 1926.

Spermogonia subepidermal, type 5. Aecia and uredinia not produced in the type. Telia subepidermal in origin, erumpent; spores 3-celled by vertical septa, thus radially arranged, deeply pigmented, opaque, subtended by 3 colorless, hygroscopic cysts surmounting a simple pedicel, germ pore 1 in each cell, germination unknown but basidium undoubtedly external.

TYPE: Cystomyces costaricensis Syd.

Only one species is known. It is microcyclic on an unidentified legume in Costa Rica. Cystomyces is similar to Ravenelia but has only 3-celled teliospores and a simple pedicel. The pedicel attaches to the cysts which are, therefore, similar to the apical cells of Dicheirinia.

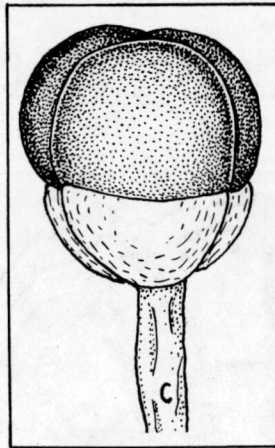

C. costaricensis Syd.;
a teliospore.

97. SPUMULA Mains, Mycologia 27:638. 1935.

Spermogonia subcuticular, type 7. Aecia subepidermal in origin, erumpent, with peridium, aecidioid; spores catenulate, verrucose. Uredinia subepidermal in origin, erumpent; spores borne singly on pedicels, echinulate, pores obscure. Telia subepidermal in origin, erumpent; spores 1-celled, laterally united in 3- to several-celled discoid, pigmented heads subtended by colorless, hygroscopic cysts, germ pore 1 in each cell, pedicel simple, basidium external.

TYPE: Spumula quadrifida Mains.

Spumula differs from Ravenelia because of the simple pedicel. The type and two other species are similar, with four spores per head. S. heteromorpha (Doidge) Thir. and S. heteromorpha J.W. Baxt. have more complicated heads, the former with mostly two to ten spores and the latter much more variable but to 20. S. heteromorpha is difficult to place. The teliospore heads have simple pedicels but the cysts seem not to be hygroscopic and not always of the same number as the spores in a head. It resembles Dicheirinia as much as it does Spumula. All species are in the Leguminosae.

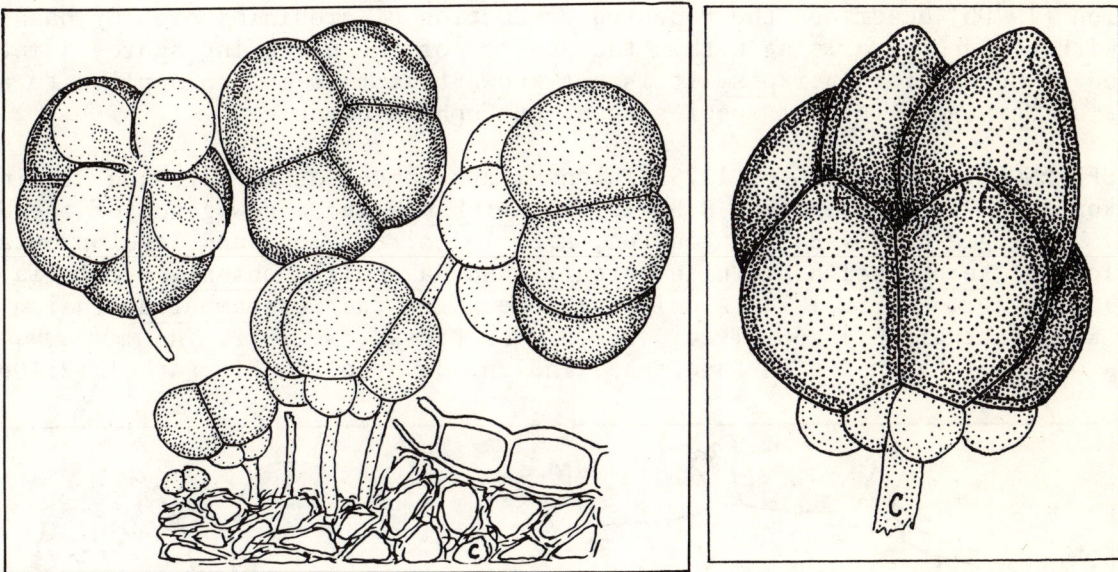

S. quadrifida Mains; teliospores (left). S. heteromorpha J.W. Baxt.; a teliospore (right).

98. KERNKAMPELLA Rajendren, Mycologia 62:839. 1970.

Spermogonia subcuticular, type 7. Aecia subepidermal in origin, erumpent, with peridium (Tyagi & Prasad) or without, uredinoid (Rajendren by implication); spores not described by either. Uredinia subepidermal in origin, erumpent, with paraphyses; spores borne singly on pedicels, echinulate, pores mostly equatorial. Telia subepidermal in origin, erumpent; spores 1-celled, each cell with a central appendage, strongly adherent laterally, forming discoid heads subtended by a patelliform layer of cells below which are hygroscopic cysts, germ pore 1 in each cell, pedicel compound, basidium external.

TYPE: Kernkampella breyniae-patentis (Mund. & Thir.) Rajen. but perhaps actually K. kirganellae (Mund. & Thir.) Laund.).

This genus of eight species is distinguished from Ravenelia because of the patelliform layer of cells between the spores and the cysts. Whether this structure warrents segregation of a genus or of a section of Ravenelia is a matter of opinion. Laundon accepts the genus; Tyagi does not. We favor retaining Kernkampella in the absence of a modern monograph of Ravenelia and its possible segregates. Rajendren (1970b) describes the repeated production of urediniospores by basal cells with new spores arising within the pedicel of the preceding spore, a method reported by Hennen for Intrapes. It is not known that this method applies to all species of Kernkampella. All species occur on Euphorbiaceae.

REFERENCES: Laundon, G.F. 1975. Taxonomy and nomenclature notes on Uredinales. Mycotaxon 3:133-161. Rajendren, R.B. 1970a. Kernkampella: a new genus of the Uredinales. Mycologia 62:837-843. Rajendren, R.B. 1970b. Cytology and developmental morphology of Kernkampella breyniae-patentis and Ravenelia hobsoni. Mycologia 62: 1112-1121. Tyagi, R.N.S. 1974. A critical account of the Kernkampella. Indian J. Mycol. & Pl. Pathol. 3:63-66. Tyagi, R.N.S. and Prasad, N. 1972. The monographic studies on genus Ravenelia in Rajasthan. Indian. J. Mycol. & Pl. Pathol. 2:108-127.

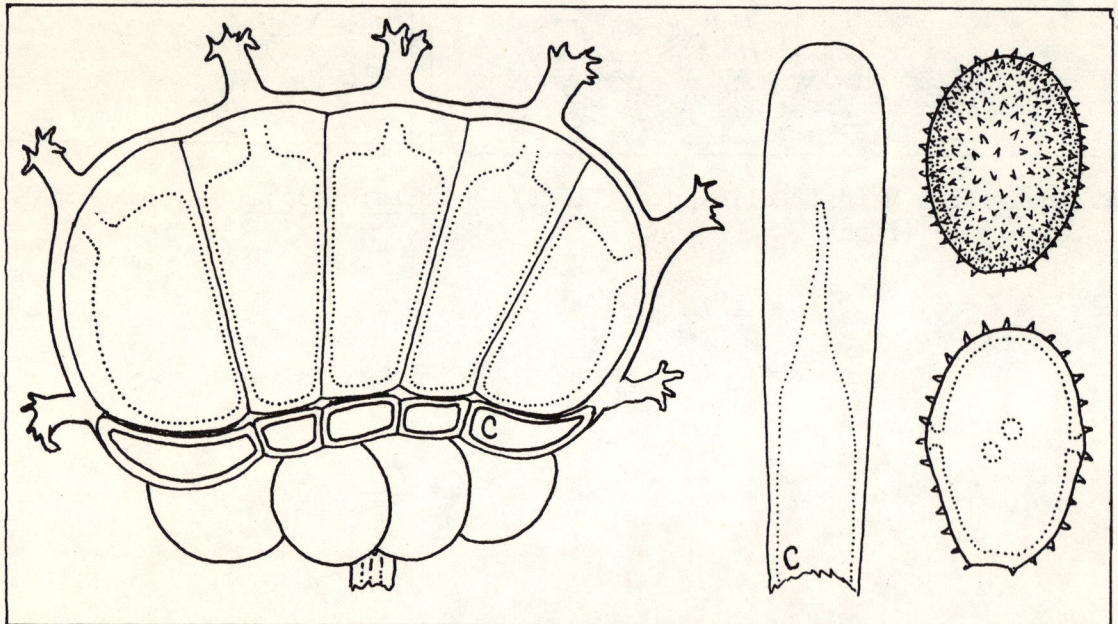

K. appendiculata (Lagh. & Diet.) Laund.; section of a teliospore head, one paraphysis, and urediniospores.

99. RAVENELIA Berkeley, Gard. Chron. 1853:132. 1853.

Spermoognia mostly subcuticular, type 7, in a few species subepidermal, type 5. Aecia subepidermal or sometimes subcuticular, erumpent, mostly uredinoid with spores pedicellate or in a few species aecidioid with catenulate spores. Uredinia mostly subepidermal but sometimes subcuticular, often paraphysate, erumpent; spores borne singly on pedicels, mostly echinulate, pores various. Telia subepidermal or sometimes subcuticular in origin, erumpent; spores typically 1-celled, in a few species 2-celled by horizontal septum, strongly adherent laterally in pedicellate, discoid heads, pigmented, surface smooth or sculptured, subtended by colorless, hygroscopic cysts, pedicels fasciculate, germ pore 1 in each spore, basidium external.

LECTOTYPE: Ravenelia glandulosa Berk. & Curt.

Ravenelia is the third largest genus with probably 150-170 species distributed world-wide in warm climates and predominantly on Leguminosae. Most species, perhaps all, reported on Euphorbiaceae belong in Kernkampella. R. atridis Syd. is on Grewia of the Tiliaceae and R. sarmentoi Lindq. is on Bulnesia (Zygophyllaceae). All species are autoecious. Microcyclic species are known. There is no recent monograph.

REFERENCES: Cummins, G.B. 1978. Rust fungi on legumes and composites in North America. Univ. Arizona Press. 424 pp. Doidge, E.M. 1927. South African rust fungi. Bothalia 2(1a):1-228. Lindquist, J.C. 1954. Las especies Argentinas de Ravenelia. Rev. Fac. Agron. 30:103-128. Sydow, P. & H. 1915. Monographia Uredinearum Vol. III. pp. 224-310. Tyagi, R.N.S. and Prasad, N. 1972. The monographic studies on genus Ravenelia occurring in Rajasthan. Indian. J. Mycol. & Pl. Pathol. 2:108-127.

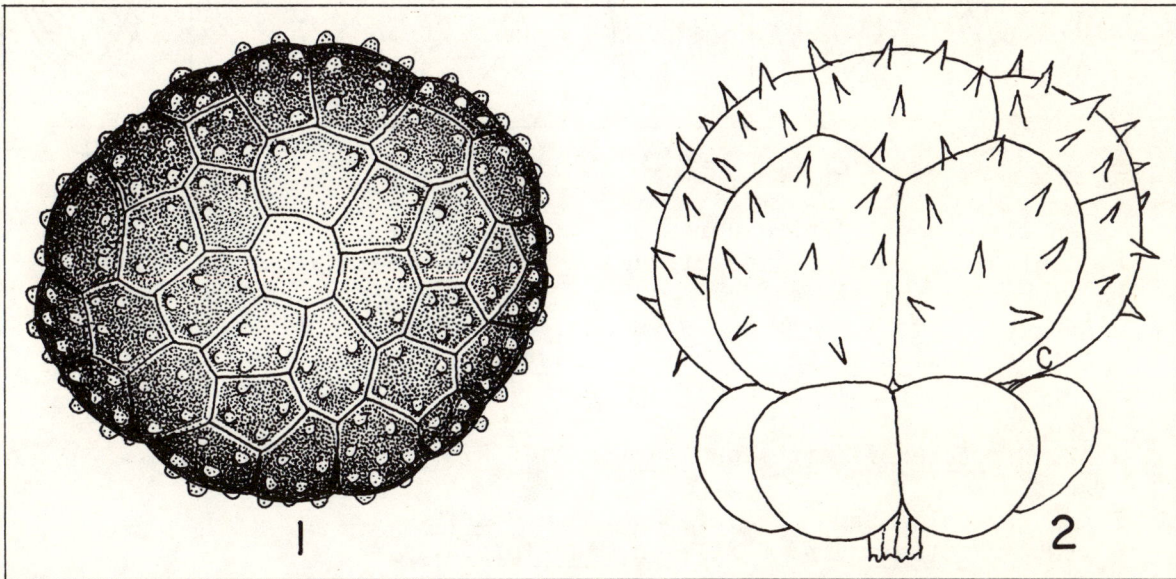

1. R. fragrans Long var. fragrans; dorsal view of a teliospore head.
2. R. echinata Lager. & Diet. var. ectypa (Arth. & Holw.) Cumm.; side view of a teliospore head showing cysts and pedicel.

Additional figures next page.

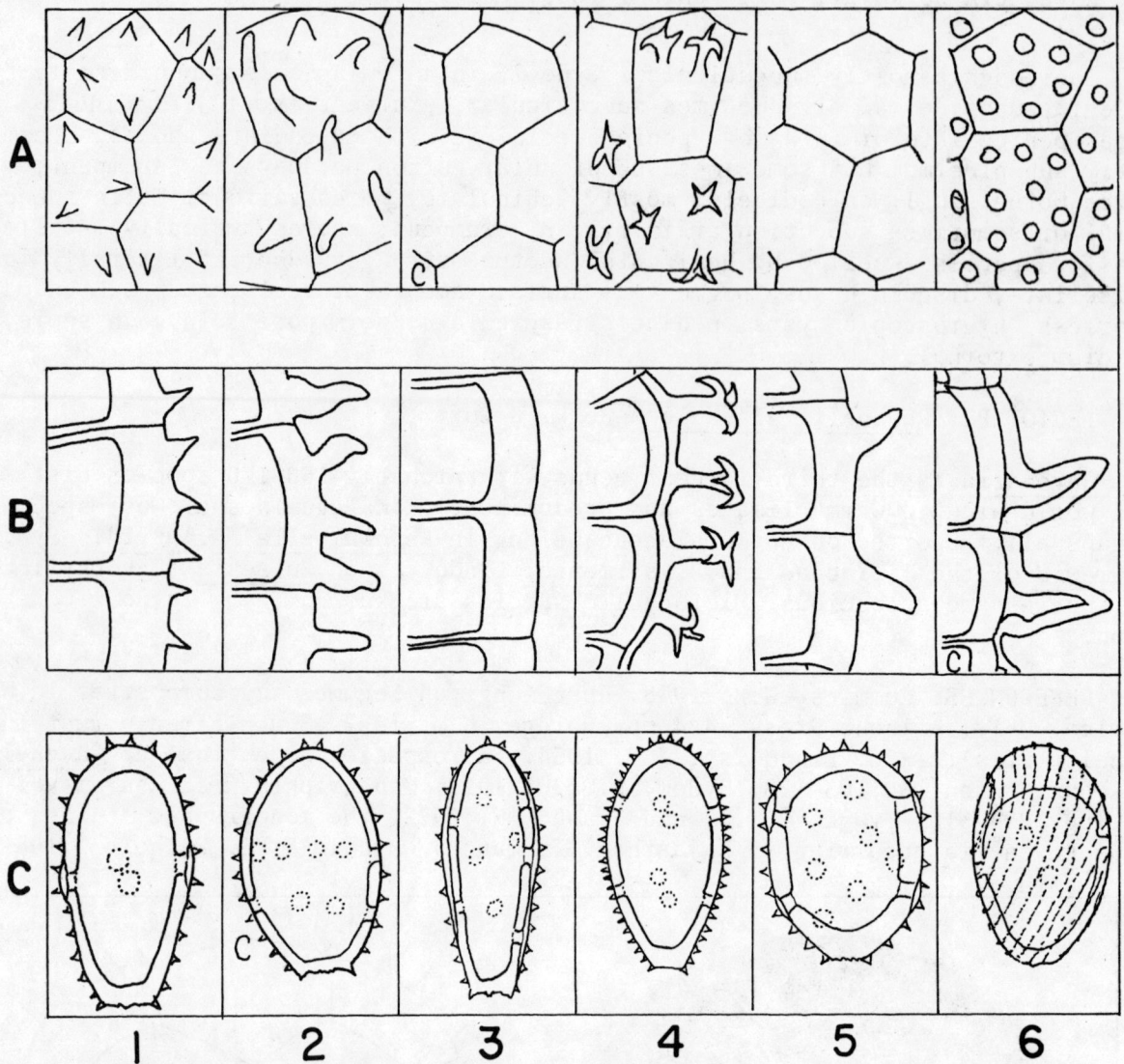

Ravenelia species; teliospores and urediniospores.
A, central cells of teliospore head; B, marginal cells of teliospore head; C, urediniospores.
1 ABC, R. hermosa Cumm. & J.W. Baxt., on Acacia.
2 ABC, R. mexicana Tranz., on Calliandra.
3 AB, R. holwayi Diet., on Prosopis; 3 C, R. cumminsii J.W. Baxt., on Acacia.
4 AB, R. stevensii Arth. on Acacia; 4 C, R. spegazziniana Lindq., on Acacia.
5 ABC, R. bella Cumm. & J.W. Baxt., on Cassia.
6 ABC, R. corbula J.W. Baxt., on Caesalpinia.

100. HAPALOPHRAGMIUM H. & P. Sydow, Hedwigia (Beibl.) 40:64. 1901.

Spermogonia subcuticular, type 7 or less commonly subepidermal, type 5. Aecia subepidermal in origin, erumpent, mostly paraphysate, uredinoid; spores borne singly on pedicels, echinulate. Uredinia similar to aecia; spores borne singly on pedicels, echinulate, pores equatorial. Telia subepidermal in origin, erumpent; spores borne singly on simple pedicels, triquetrously 3-celled with 2 cells basal and 1 above, wall pigmented, smooth or sculptured, germ pore 1 in each cell, basidium external.

TYPE: Hapalophragmium derridis Syd.

This is a genus of 15 species, all on legumes in tropical Asia and Africa.:. all are autoecious, and most are macrocyclic. We treat Hapalophrgmiopsis Thir. as synonymous, as does Monoson, but Laundon cites possible reasons for keeping it separate, as did Thirumalachar.

REFERENCES: Laundon, G.F. 1975. Taxonomy and nomenclature notes on Uredinales. Mycotaxon 3:133-161. Monoson, H.L. 1977. A synopsis of the genus Hapalophragmium (Uredinales). Mycologia 69:21-33. Thirumalachar, M.J. 1950. Some noteworthy rusts. III. Mycologia 42:224-232.

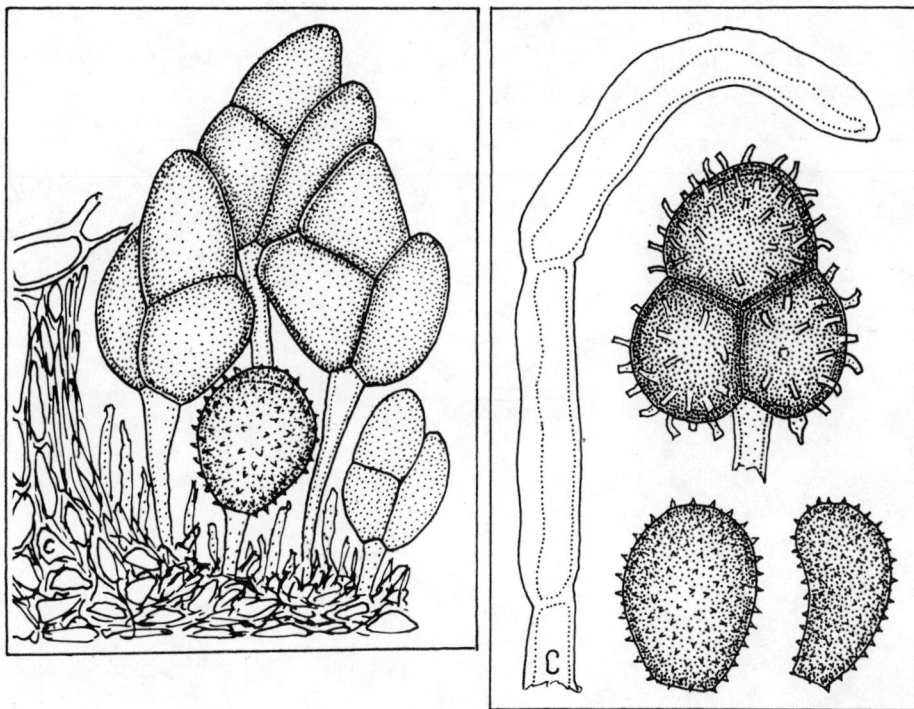

H. derridis Syd.; a sorus with teliospores and one urediniospore (left). H. ornatum Cumm.; a paraphysis, a teliospore, and urediniospores (right).

101. TRIPHRAGMIUM Link in Linnaeus Species Plantarum 6:84. 1825.

Spermogonia subcuticular, type 11. Aecia subepidermal in origin, erumpent; spores borne singly on pedicels, echinulate. Uredinia subepidermal in origin, erumpent, with paraphyses; spores borne singly on pedicels, echinulate, pores obscure. Telia subepidermal in origin, erumpent; spores borne singly on pedicels, triquetrously 3-celled with the basal pedicellate cell surmounted by 2 cells, wall pigmented, sculptured, especially around the pores, germ pore 1 in each cell, basidium external.

TYPE: Triphragmium ulmariae (DC.) Link (Puccinia ulmariae DC.).

Four species parasitize the Rosaceae in the Northern Hemisphere and two are on Leguminosae. The genus differs form Nyssopsora and Triphragmiopsis in having one pore per cell and from Hapalophragmium because the odd cell is basal. Triphragmiopsis and Triphragmium have teliospores with similar surface sculpture which is unlike that of Nyssopsora.
Triactella Syd. (type; Triphragmium pulchrum Racib. on Derris, despite the host being Leguminosae rather than Rosaceae, is not now separable by sustantial characters and is treated here as a synonym. Its spermogonia and aecia are not known. Should the spermogonia prove to be type 7, and this is probable, then a review of this treatment may be desirable.

REFERENCE: Monoson, H.L. 1974. The species of Triphragmium, Nyssopsora, and Triphragmiopsis. Mycopathol. Mycol. Appl. 52:115-131.

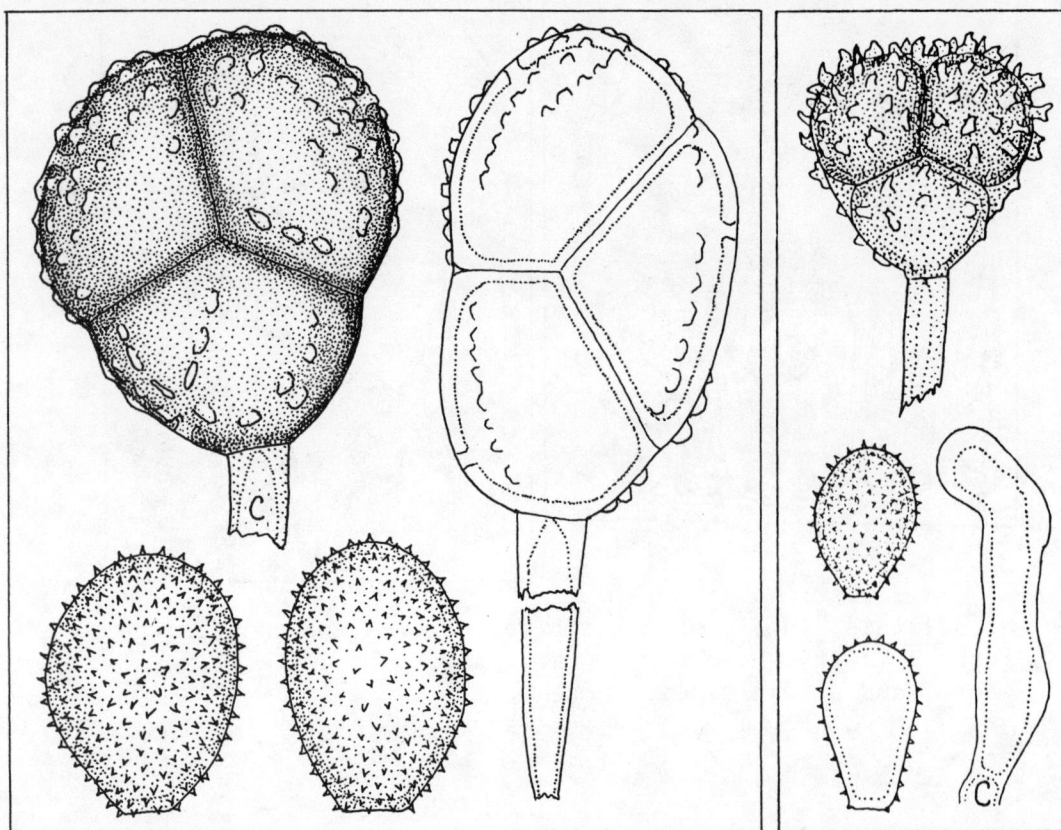

T. ulmariae (DC.) Link; teliospores and urediniospores (left).
T. pulchrum Racib.; a teliospore, a paraphysis, and urediniospores.

102. TRIPHRAGMIOPSIS Naumov, Bull. Soc. Mycol. France 30:78. 1914.

Spermogonia unknown. Aecia subepidermal in origin, erumpent, with peridium; spores catenulate, echinulate. Uredinia unknown. Telia subepidermal in origin, erumpent; spores borne singly on pedicels, triquetrously 3-celled with a basal pedicellate cell surmounted by 2 cells, wall pigmented and verrucose, germ pores 2 in each cell, basidium external.

TYPE: Triphragmiopsis jeffersoniae Naum.

There are two species, the type on Berberidaceae in the western USSR and Korea, and T. isopyri (Moug. & Nestl) Tranz. on Ranunculaceae in Europe. The genus differs from Triphragmium in having two germ pores per teliospore cell and from Nyssopsora because of different wall sculpture. The appearance of the teliospores is more that of Triphragmium than of Nyssopsora. The hosts are unlike those of either of these genera. The consequence is that Triphragmiopsis has both been accepted and submerged. Recently it has been accepted by Monoson and Azbukina and submerged in Nyssopsora by Majewski. Henderson suggests, because of spore ornamentation, relationship with the Phragmideae.

REFERENCES: Azbukina, Z.M. 1974. Rust fungi of the Soviet Far East (in Russian). Moscow. 527. pp. Henderson, D.M. 1973. The rust fungus Nyssopsora and its relations. Notes Roy. Bot. Gard. Edinb. 32: 217-221. Majewski, T. 1977. Uredinales I, in Mycota (Grzyby) Tom IX. Warsaw-Krakow. 394 pp. Monoson, H.L. 1974. The species of Triphragmium, Nyssopsora, and Triphragmiopsis. Mycopathol. Mycol. Appl. 52:115-131.

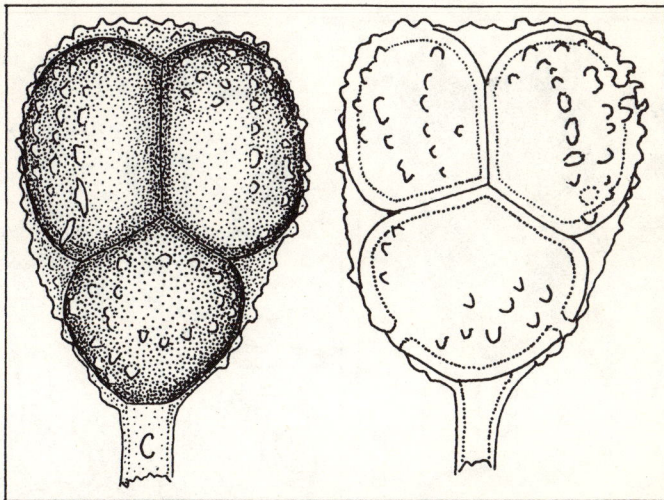

T. jeffersoniae Naum.; teliospores.

103. NYSSOPSORA Arthur, Rés. Sci. Congr. Internat. Vienne p. 342. 1906.

Spermogonia and aecia unknown. Uredinia subepidermal in origin, erumpent; spores borne singly on pedicels, echinulate. Telia subepidermal in origin, erumpent; spores borne singly on pedicels, triquetrously 3-celled with a basal, pedicellate cell surmounted by 2 cells, wall pigmented and bearing conspicuous spines which are often apically branched, germ pores 2 in each cell, basidium external.

TYPE: Nyssopsora echinata (Lév.) Arth. (Triphragmium echinatum Lév.).

There are nine species, four on Araliaceae, two on Sapindaceae, and one each on Meliaceae, Pittosporaceae, and Umbelliferae distributed in Asia, Australia,Europe, and North America. Information about life cycles is scant. Aecia are unknown or uncertain; some are microcyclic. The strongly spinose teliospores are the conspicuous feature and separate the genus from Triphragmium (with one germ pore per cell) and Triphragmiopsis (with two pores per cell). Henderson states that the sculpture of the teliospore wall is different from Triphragmium (and doubtless also from Triphragmiopsis).

REFERENCES: Luetjeharms, W.J. 1937. Vermischte Mycologische Notizen I. Ueber die Gatung Nyssopsora. Blumea, Suppl. 1:142-161. Henderson, D.M. 1969. Studies in the morphology of fungal spores I: the teliospores of Puccinia prostii and Nyssopsora echinata. Notes Roy. Bot. Gard. Edinb. 29:373-375. Henderson, D.M. 1973. The rust genus Nyssopsora and its host relations. Notes Roy. Bot. Gard. Edinb. 32:217-221. Monoson, H.L. 1974. The species of Triphragmium, Nyssopsora, and Triphragmiopsis. Mycopathol. Mycol. Appl. 52:115-131.

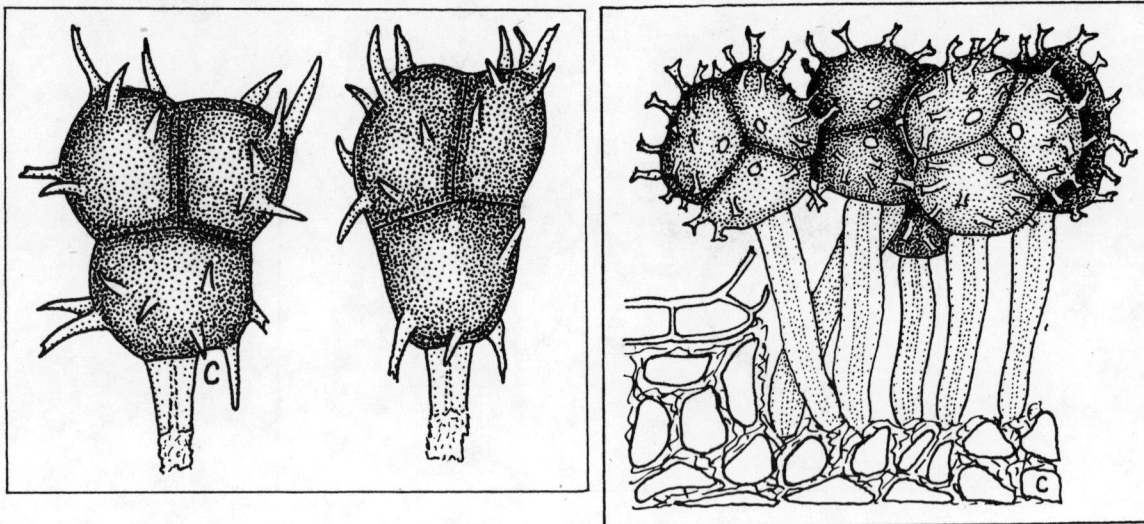

N. echinata (Lév.) Arth.; teliospores (left). N. clavellosa (Berk.) Arth.; teliospores (right).

104. SPHAEROPHRAGMIUM Magnus, Ber. Dtsch. Bot. Ges. 9:121. 1891.

Spermogonia type 5. Aecia subepidermal in origin, erumpent, with peridium; spores catenulate, verrucose. Uredinia subepidermal in origin, erumpent; spores borne singly on pedicels, echinulate, pores equatorial, the spores often asymmetrical and reniform, mostly with peripheral paraphyses. Telia subepidermal in origin, erumpent; spores borne singly on pedicels, 4- to several-celled by vertical and horizontal septa, usually more or less globoid, wall pigmented, with simple or often apically furcate spines or projections, germ pores obscure, uncertain, basidium external.

TYPE: Sphaerophragmium acaciae (Cooke) Magn. (Triphragmium acaciae Cooke).

There are 18 species; most are on legumes, and they occur circumglobally in warm to tropical areas.

REFERENCES: Hiremuth, R.V. and Pavgi, M.S. 1974. Development of the telium in Sphaerophragmium acaciae. Norweg. J. Bot. 21:17-21. Monoson, H.L. 1974. The genus Sphaerophragmium. Mycologia 66:791-802.

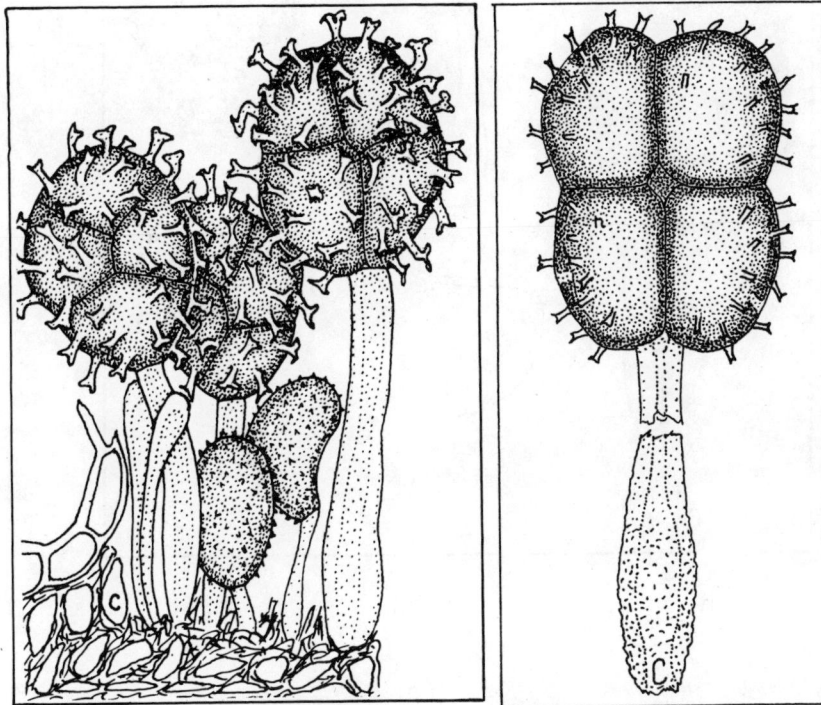

S. acaciae (Cooke) Magn.; paraphyses, urediniospores, and teliospores (left). S. artabotrydis Doidge; one teliospore (right).

105. CUMMINSINA Petrak, Sydowia 9:474. 1955.

    Spermogonia and aecia unknown. Uredinia subepidermal in origin, erumpent,
with peripheral paraphyses; spores borne singly on pedicels, echinulate. Telia
subepidermal in origin, erumpent; spores consisting of chains of laterally
adherent cells to form a club-shape, pigmented head, pedicel simple basally but
with apical cells equal in number to the chains of cells, germ pores not seen,
basidium undoubtedly external but germination unknown.

    TYPE: Cumminsina clavispora Petr.

    The only known species occurs on Grewia of the Tiliaceae in Africa. The
genus has the features of Anthomycetella but produces more complex (often more so
than illustrated) spore heads which appear like a bundle of Phragmidium teliospores or a giant Alternaria spore. The relationship of Cumminsina is obscure.

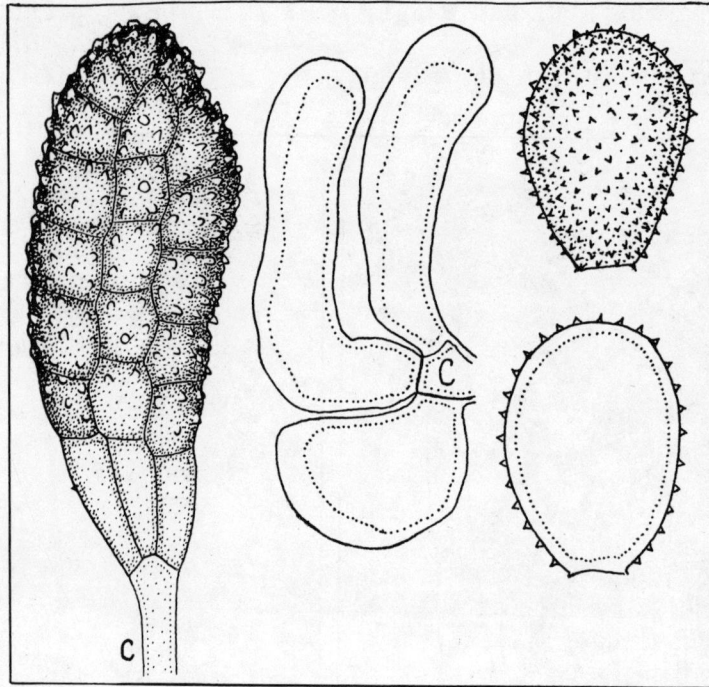

C. clavispora Petr.; paraphyses, urediniospores
and a teliospore.

USEFUL GENERAL REFERENCES
(excluding textbooks)

Arthur J.C. The Plant Rusts. New York: Wiley & Sons; 1929.

Azbukina, Z.M. Rust Fungi of the Soviet Far East. "Preface" (in Russian). Moscow: Akad. Nauk CCCP; 1974.

Dietel, P. Reihe Uredinales (in German). In Engler & Prantl Nat. Pflanzenfam. 6: 24-35. 1928.

Hiratsuka, N. Uredinological Studies (in Japanese). Tokto: Kasai Publ. Co.; 1955.

León-Gallegos, H.; Cummins, G.B. Uredinales (Royas) de México. "Introduccion" (in Spanish) Culiacán, Sin.: 1981.

Lindquist, J.C. Royas de la Republica Argentina y Zonas Limitrofes. Chapt. I (in Spanish). Coleccion Cient. Tomo XX; 1982.

Littlefield, L.J. Biology of the Rust Fungi. Ames: Iowa State Univ. Press; 1981.

Littlefield, L.J.; Heath, M.C. Ultrastructure of Rust Fungi. New York-London: Academic Press; 1979.

Scott, N.J.; Chakravorty, A.K. (Eds.). The Rust Fungi. New York-London: Academic Press; 1982.

DESCRIPTIVE MANUALS

Arthur, J.C. Manual of the Rusts in United States and Canada. Lafayette, IN: Purdue Res. Found.; 1934.

Azbukina, Z.M. Rust Fungi of the Soviet Far East (in Russian). Moscow: Akad. Nauk CCCP; 1974.

Bakshi, B.K.; Singh, S. Rusts on Indian Forest Trees. Delhi: Govt. India Press; 1967.

Buritica, P.; Hennen, J.F. Pucciniosireae (Uredinales, Pucciniaceae). Flora Neotropica. Monogr. No. 24; 1980.

Cummins, G.B. The Rust Fungi of Composites and Legumes in North America. Tucson, AZ: Univ. Arizona Press; 1978.

Cummins, G.B. The Rust Fungi of Cereals, Grasses and Bamboos. New York-Heidelberg-Berlin: Springer-Verlag; 1971.

Cunningham, G.H. The Rust Fungi of New Zealand. Dunedin: McIndoe; 1931.

Doidge, E.M. Preliminary Study of the South African Rust Fungi. Bothalia 2(1):1-228; 1927; 2(2):473-474; 1928; 3(4):487-512; 1939; 4(1):229-236; 4(4):895-937; 1948.

Gäumann, E. Die Rostpilze Mitteleuropas. Bern: Büchler; 1959.

Gjaerum, H. Nordens Rustsopper. Oslo: Fungiflora; 1974.

Hiratsuka, N. A Monograph of the Pucciniastreae. Japan: Mem. Tottori Agr. Col.; 1936; revised Kasai Publ. Co.; 1958.

Ito, S. Mycological Flora of Japan. Vol. II (in Japanese). Tokyo; 1950.

Kuprevich, V.F.; Tranzschel, V.T. Cryptogamic Plants of the USSR. Fungi (1), Rust Fungi No. 1, fam. Melampsoraceae. Moscow-Leningrad; 1957. Translation publ. U. S. Dept. Agr. and Natl. Sci. Found.; 1970.

Kuprevich, V.F.; Ul'yanischev, V.I. Key to Rust Fungi of the USSR. Melampsoraceae and some genera of Pucciniaceae (in Russian). Minsk; 1975.

León-Gallegos, H.; Cummins, G.B. Uredinales (Royas) de México. Vol. I, II. Culiacán, Sin.: 1981.

Lindquist, J.C. Royas de la Republica Argentina y Zonas Limitrofes. Coleccion Cient. Tomo XX; 1982.

Majewski, T. Grzby (Mycota). Tom IX, XI (Uredinales). Warszawa-Krakow; 1977, 1979.

McAlpine, D. The Rusts of Australia. Melbourne: Brain; 1906.

Savulescu, T.R. Monografia Uredinalelor din Republica Populara Romana. Vol. II. Bucurest; 1953.

Sydow, P.; Sydow, H. Monographia Uredinearum. Vol. I-4. Leipsig: Borntraeger; 1902-1924. (World coverage).

Ul'yanishchev, V.I. Key to the Rust Fungi of the USSR. Pt. 2. (in Russian). Leningrad; 1978.

Viégas, A.P. Alguns Fungos do Brazil IV, Uredinales. Bragantia 5(1); 1945.

Wilson, M.; Henderson, D.M. British Rust Fungi. Cambridge; Cambridge University Press; 1966

Ziller, W.G. The Tree Rusts of Western Canada. Can. For. Serv. Publ. No. 1329; Victoria; 1974.

CHANGES OF NOMENCLATURE

Arthuriomyces (genus No. 78) gen. nov. (Gymnoconia sense of Lagerheim, not type).
Type: Arthuriomyces peckianus (Howe in Peck) Cumm. & Y. Hirat., comb. nov.
(Puccinia peckiana Howe in Peck, Ann. Rept. N. Y. State Mus. 23:57. 1872).

Atelocauda (genus No. 51) bicincta (McAlp.) Cumm. & Y. Hirat., comb. nov. (Uromyces bicinctus McAlp., Rusts of Australia p. 93. 1906).

Atelocauda digitata (Wint.) Cumm. & Y. Hirat., comb. nov. (Uromyces digitatus
Wint., Rev. Mycol. 8:209. 1886).

Atelocauda koae (Arth.) Cumm. & Y. Hirat., comb. nov. (Uromyces koae Arth., in
Stevens, Bernice P. Bishop Mus. Bull. 19:118. 1925).

Chrysocelis (genus No. 17) geophilicola (Yen) Cumm. & Y. Hirat. comb. nov. (Stomatisora geophilicola Yen, Rev. Mycol. 35:332. 1971).

Diorchidium (genus No. 93) copaifera (Syd.) Cumm. & Y. Hirat., comb. nov. (Sphenospora copaifera Syd., Monogr. Ured. 4:584. 1924).

Edythea (genus No. 58) palmaea (Hennen & Ono) Cumm. & Y. Hirat. comb. nov. (Cerradoa palmaea Hennen & Ono, Mycologia 70:570. 1978).

Frommeëlla (genus No. 83) genus nov. Cumm. & Y. Hirat. (Frommea sense of Arthur,
not type).
Type: Frommeëlla tormentillae (Fckl.) Cumm. & Y. Hirat. comb. nov.)Phragmid-
ium tormentillae Fckl. Jahrb. Nass. Ver. Nat. 23-24:46. 1870).

Gerwasia (genus No. 45) standleyi (Cumm.) Cumm. & Y. Hirat., comb. nov. (Mainsia
standleyi Cumm., Bull. Torrey Bot. Club 70:71. 1943).

Maravalia (genus No. 49) kevorkianii (Cumm.) Cumm. & Y. Hirat. comb. nov. (Scopella kevorkianii Cumm., Bull. Torrey Bot. Club 77:206. 1950).

Olivea (genus No. 20) fimbriata (Mains) Cumm. & Y. Hirat., comb. nov. (Tegillum
fimbriatum Mains, Bull. Torrey Bot. Club 67:707. 1940).

Porotenus (genus No. 68) depallens (Arth. & Holw.) Cumm. & Y. Hirat., comb. nov.
(Puccinia depallens Arth. & Holw. in Arthur, Mycologia 10:139. 1918).

Porotenus elatipes (Arth. & Holw.) Cumm. & Y. Hirat., comb nov. (Puccinia elatipes
Arth. & Holway in Arthur, Mycologia 10:133. 1918).

Porotenus permagnus (Arth. & Holw.) Cumm. & Y. Hirat., comb. nov. (Puccinia permagna Arth. & Holw. in Arthur, Mycologia 10:134. 1918).

Sorataea (genus No. 67) cerasi (Cast.) Cumm. & Y. Hirat., comb. nov. (Puccinia
cerasi Cast., Obs. Pl. Acot. Fam. Ured. 1:13. 1842; Leucotelium cerasi (Cast.)
Tranz.).

Sorataea holwayi (H.S. Jack.) Cumm. & Y. Hirat., comb. nov. (Mimema holwayi H.S.
Jack., Mycologia 23:338. 1931).

Sorataea padi (Tranz.) Cumm. & Y. Hirat., comb. nov. (Leucotelium padi Tranz.,
Sov. Bot. 4:84. 1935).

Sorataea pruni-persicae (Hori) Cumm. & Y. Hirat., comb. nov. (Puccinia pruni-
persicae Hori, Phytopathology 2:344. 1912; Leucotelium pruni-persicae (Hori)
Tranz.).

GLOSSARY

AECIAL INITIALS: Egg cells, female gametes; the site of initiation of the dikaryo-phase (Fig. 2).

AECIDIOID: Applied to aecia, uredinia, or telia that resemble the anamorphic genus Aecidium, i. e., have a peridium and catenulate spores (Fig. 3A).

AECIOSPORE: A spore produced as a result of mating and which germinates vegeta-tively; applies to spores produced by the anamorphic genera Aecidium, Caeoma, Elateraecium, Peridermium, Roestelia, and Uraecium.

AECIUM: A sorus initiated on a haploid thallus together with spermogonia, develop-ing binucleate spores following transfer of sperms to eggs, and producing vegetative mycelium, not a basidium; a generalized term that applies to any of the anamorphic genera Aecidium, Caeoma, Elateraecium, Peridermium, Roestelia, and Uraecium (Fig. 2, 3).

AMPHISPORE: A specialized urediniospore that has a wall that is thicker and usu-ally more deeply pigmented than the ordinary urediniospore of the species (see genus No. 3).

ANAMORPH: An "Imperfect" state, i. e., not the basidial state.

APICAL CELLS: Cells of the upper part of a pedicel and to which spores are at-tached (see genera Nos. 89, 90).

AUTOECIOUS: Requiring only a single host to complete the life cycle, e. g., bean rust, Uromyces appendiculatus.

BASAL CELL: A cell that produces several spores (see genera 44, 58).

BASIDIOSPORE: A haploid spore (usually in 4's) produced by a basidium (see genus No. 3).

BASIDIUM: The structure that produces basidiospores following meiotic division of the zygote nucleus. Also see "Teliospore" (see genera Nos. 3, 5, 13, 14).

BASIDIOSORUS: A sorus comprised of basidia (see genera Nos. 13, 14; also see "Internal basidium).

BILAMINATE: Two-layered; applies to spore walls (see genera Nos. 72, 76).

CAEOMATOID: Applies to aecia, uredinia, or telia that resemble the anamorphic genus Caeoma, i. e., with catenulate spores but no peridium.

CATENULATE: Borne in lineal series or chains (see genera Nos. 22, 23).

CYSTS: Sterile, colorless, hygroscopic cells that subtend some teliospores (see genera Nos. 96, 97, 98).

DEMICYCLIC: A life cycle comprised of spermogonia, aecia, and telia but lacking uredinia; e. g., many species of Gymnosporangium.

DIORCHIDIOID: Resembling Diorchidium; two-celled teliospores with the septum more or less vertical; most often applied to species of Puccinia (see genera Nos. 57, 93).

ECHINULATE: Adorned with sharp spines or cones; characteristic of most uredinio-spores (Fig. 5A, B, C).

ENDOCYCLIC: A microcyclic life cycle in which aeciospore-like cells produce basid-ia (see genera Nos. 30E, 63E, 78E).

EXTERNAL BASIDIUM: A basidium that matures outside of a teliospore, i. e., the common kind (see genera Nos. 3, 15).

FASCICLED: Applies to pedicels composed of several adherent hyphae (see genus 99).

FLEXUOUS HYPHAE: Same as trichogynes.

GERM PORE: A small, discrete area of a spore wall through which germination oc-curs (Fig. 4, genus No. 84).

HYGROSCOPIC: Capable of absorbing water and expanding; may occur in spore walls, pedicels, and cysts (see genera Nos. 72, 76, 99).

HETEROECIOUS: Requiring two unrelated hosts for completion of the life cycle, e. g., barberry and wheat in the case of Puccinia graminis.

INTERCALARY CELLS: Sterile cells which may occur between catenulate spores (see genera Nos. 36, 39).

INTERNAL BASIDIUM: A basidium developed by septation of the protoplast of a "teliospore" (see genera Nos. 14, 43; also see "BASIDIOSORUS").

MACROCYCLIC: A life cycle comprised of aecia, uredinia, telia, and usually spermogonia; may be autoecious or heteroecious.

MICROCYCLIC: A life cycle comprised of spermogonia and telia or of telia only; obviously autoecious.

OSTIOLAR CELLS: Cells that delimit the aperture (ostiole) of peridiate uredinia, as in Pucciniastrum and Melampsoridium.

PARAPHYSES: Sterile, elongate structures within or at the edge of a sorus and usually clearly distinct from spores or pedicels (see genera Nos. 7, 18, 20).

PEDICEL: The supporting hypha of a spore.

PEDICELLATE: Having a pedicel.

PERIDERMIOID: Having the appearance of the anamorphic genus Peridermium, the aecial state of most rust fungi on conifers (Fig. 3B).

PERIDIATE: Having a peridium (Fig. 3A, B, C).

PERIDIUM: A cellular structure enclosing the sporogenous area of some sori; espeially common in aecia (Fig. 3A, B, C).

PERIPHERAL: Around the margin of a sorus, e. g., a peridium or paraphyses.

RENIFORM: Kidney-shape; applies to certain urediniospores that have an invaginated side (see genera Nos. 38, 93).

RECEPTIVE HYPHA: Same as trichogyne.

RETICULATE: Netted; applies to spore surfaces having a net-like ornamentation (Fig. 5N).

ROESTELIOID: Having the appearance of the anamorphic genus Roestelia (Fig. 3C).

SESSILE: Lacking a pedicel.

SPERMOGONIUM: The male sex organ; sometimes called pycnium. (Fig. 1, 2).

SPERMATIUM: The male gamete produced in a spermogonium (Fig. 2).

SPOROGENOUS CELLS: Cells that produce several spores; same as basal cells (see genera Nos. 44, 58).

SUPRASTOMATAL: Above stomata; applies to sori that develop outside of stomata (see genera Nos. 44, 58, 70).

SYSTEMIC: Mycelium more or less uniformly distributed within a host plant; recognized because sori develop over an entire plant or plant part, usually inciting a change in the habit of growth of the host.

TELIOSPORE: A spore that produces a basidium. Some terminologies employ probasidium or hypobasidium for the spore-like structure and metabasidium for the tube that emerges during germination and which bears the basidiospores. See also "Basidium."

TELIUM: The sorus that bears teliospores, i. e., the terminal sorus of the dikaryophase, the site of karyogamy.

TRICHOGYNE: A hyphal extension from the aecial initials (egg cells) by means of which sperm nuclei reach the egg cells; same as flexuous hyphae and receptive hyphae (Fig. 2).

UREDINIOSPORE: A spore produced on a dikaryotic mycelium and which produces dikaryotic mycelium upon germination.

UREDINIUM: A sorus initiated on a dikaryotic mycelium and whose spores germinate vegetatively, i. e., not with a basidium. This is the "repeating stage" of the rust fungi. Uredinia may have peridia, paraphyses, or neither and the spores may be catenulate, pedicellate, or sessile.

UREDINOID: Having the appearance of the anamorphic genus Uraecium; applies especially to aecial states that have the gross appearance of uredinia but are accompanied by spermogonia and which sporulate after spermatization and the establishment of the dikaryophase (Fig. 3E).

VERRUCOSE: Adorned with discrete, wart-like protuberances, but commonly applied to a variety of roughened surface intermediate between echinulate and reticulate (Fig. 5D-H).